Nothing Personal?

RGS-IBG Book Series

For further information about the series and a full list of published and forthcoming titles please visit www.rgsbookseries.com

Published

Nothing Personal?

Geographies of Governing and Activism in the British Asylum System

Nick Gill

WILEY Blackwell

Library of Congress Cataloging-in-Publication data applied for

9781444367065 [hardback]
9781444367058 [paperback]

A catalogue record for this book is available from the British Library.

Cover image: Front Cover image © Nick Gill

Set in 10/12pt Plantin by SPi Global, Pondicherry, India

The information, practices and views in this book are those of the author(s) and do not necessarily reflect the opinion of the Royal Geographical Society (with IBG).

Printed and bound in Malaysia by Vivar Printing Sdn Bhd

1   2016

For Julia

Contents

Series Editors' Preface

The RGS-IBG Book Series only publishes work of the highest international standing. Its emphasis is on distinctive new developments in human and physical geography, although it is also open to contributions from cognate disciplines whose interests overlap with those of geographers. The Series places strong emphasis on theoretically-informed and empirically-strong texts. Reflecting the vibrant and diverse theoretical and empirical agendas that characterize the contemporary discipline, contributions are expected to inform, challenge and stimulate the reader. Overall, the RGS-IBG Book Series seeks to promote scholarly publications that leave an intellectual mark and change the way readers think about particular issues, methods or theories.

For details on how to submit a proposal please visit:
www.rgsbookseries.com

David Featherstone
University of Glasgow, UK

Tim Allott
University of Manchester, UK
RGS-IBG Book Series Editors

List of Figures

Acronyms

BBC	British Broadcasting Corporation
BME	British Minority Ethnic
BNP	British National Party
BRIAF	Bristol Refugee Inter Agency Forum
CAB, CABx	Citizens Advice Bureaux
Cedars	Compassion, Empathy, Dignity, Approachability, Respect and Support
DCO	Detention Custody Officer
DFT	Detained Fast Track
DHSS	Department of Health and Social Security
ESRC	Economic and Social Research Council
EVF	English Volunteer Force
FAS	Failed Asylum Seeker
G4S	Group 4 Securicor
IND	Immigration and Nationality Directorate
NASS	National Asylum Support Service
NHS	National Health Service
PCS	Public and Commercial Services (union)
PO	Presenting Officer
RATs	Regional Asylum Teams
SLC	South London Citizens
UAF	Unite Against Fascism
UKBA	United Kingdom Border Agency
UNHCR	United Nations High Commission for Refugees

Acknowledgements

I am grateful to lots of people who have helped in the writing of Nothing Personal? I thank the activists of Bristol for their inspiration. The participants in the research, who have been generous enough to give their time to the project, also deserve a special mention. Others include, in particular, the co-investigators and researchers who I have had the pleasure of working alongside: Jennifer Allsopp, Mary Bosworth, Andrew Burridge, Deirdre Conlon, Abigail Grace, Melanie Griffiths, Alexandra Hall, Dominique Moran, Ceri Oeppen, Natalia Paszkiewicz, Rebecca Rotter and Imogen Tyler. I have also derived great benefit from conversations with my students working in this area, including Shoker Abobeker, Emma Marshall, Patrycja Pinkowska and Amanda Schmid-Scott. Others who have provided encouragement and suggestions include John Allen, Clive Barnett, Keith Bassett, Sean Carter, Paul Cloke, Matt Finn, Wendy Larner, Krithika Srinivasan, Adam Tickell, Gordon Walker and Andrew Williams. I am grateful in particular to Jonathan Darling and the reviewers for constructive, patient and supportive comments through the writing process and to Neil Coe for fantastic editorial support. There are certainly others, too numerous to mention, who have been an invaluable support, including not least the staff at Wiley-Blackwell and Terry Halliday for work on the index. I acknowledge the financial support of the Economic and Social Research Council (grant numbers PTA-030-2003-01643, RES-000-22-3928-A, ES/J023426/1 and ES/J021814/1). Finally, a huge thank you to my brother Mat for lots of advice and encouragement, my parents, and to my wonderful family.

Chapter One
Introduction

Nothing personal, it's just business: this is the new Satan of liquid modernity.
Bauman and Donskis (2013, p. 10)

Migrant Deaths

In 2013 an unannounced inspection of Harmondsworth Immigration
Removal Centre revealed worrying instances of neglect. Harmondsworth
is a British secure facility near London that incarcerates refused asylum
seekers prior to their deportation. The inspection, undertaken by Her
Majesty's Chief Inspector of Prisons, reported that 'on at least two occasions, elderly, vulnerable and incapacitated detainees, one of whom was
terminally ill, were needlessly handcuffed in an excessive and unacceptable
manner... These men were so ill that one died shortly after his handcuffs
were removed and the other, an 84 year-old-man, died while still in
restraints' (HM Chief Inspector of Prisons, 2014, p. 5). Staff had ignored a
doctor's report declaring the 84-year-old, Alois Dvorzac, unfit for detention
and in need of medical care. 'These are shocking cases where a sense of
humanity was lost' the report continued, '[n]either had been in any way
resistant or posed any current specific individual risk' (HM Chief Inspector
of Prisons 2014, p. 13). Harmondsworth has the capacity to hold 615
detainees, making it the largest detention centre in Europe. It holds men

Nothing Personal?: Geographies of Governing and Activism in the British Asylum System,
First Edition. Nick Gill.
© 2016 John Wiley & Sons, Ltd. Published 2016 by John Wiley & Sons, Ltd.

only and the security in various wings is comparable to a high security prison. The report concluded that the centre displayed, 'inadequate focus on the needs of the most vulnerable detainees, including elderly and sick men, those at risk of self harm through food refusal, and other people whose physical or mental health conditions made them potentially unfit for detention' (HM Chief Inspector of Prisons, 2014, p. 5).

Mr Dvorzac's specific case is not an isolated phenomenon. Deaths in immigration detention are part of a global pattern of migrant deaths that occur as a result of the combination of bureaucratic ineptitude, the desperation of migrants and the strengthening of border controls. What is more, is not just asylum seekers who face risks.[1] For example, 58 Chinese stowaways who had suffocated in a container en route to the UK to work were discovered in Dover in 2001, together with just two survivors, almost suffocated amidst the putrid smell of rotting corpses (Hyland, 2000). The migrants had travelled from the southern Chinese province of Fujian on the Taiwan Strait and would have paid around £15,000 to get to Britain, most likely travelling on the strength of a deposit and facing the rest of the debt upon their arrival.[2] Although widespread consternation was expressed at the time, no fundamental alterations were made to the border policies and control practices that are at least partly responsible for the high risks they took. Another 23 Chinese migrants died picking cockles on the sands of Morecambe Bay in Lancashire, United Kingdom, in 2004. They were employed illegally, paid well below the minimum wage, and were sent to work in dangerous conditions without safety equipment or the ability to call for help. When the tide suddenly came in they were swept out to sea and suffered 'death in a cold, strange land' (BBC, 2006a). Although their deaths prompted the adoption of the Gangmaster (Licensing) Act (GLA) 2004, there 'is little direct evidence to suggest that the GLA has reduced worker exploitation, including long hours, lack of holiday and/or sick pay, unfair deductions, poor-quality tied housing, and restrictive contracts' (Strauss, 2013, p. 190). More recently, one man died and another 34 others were found suffering from dehydration and hypothermia, in a shipping container in Tilbury Docks, Essex, in August 2014. In this case the group were Afghan Sikhs who were intending to claim asylum, and included 13 children; they had been trapped inside the container for at least 12 hours.

The moral claim made by asylum seekers is seen as different from that made by economic migrants even though both often experience hardship, uncertainty and discomfort. Asylum seekers are invoking their right to safety from persecution rather than their right to work. As such they do not offend the sensibilities of those who are concerned about 'British jobs for British workers' in quite the same way as economic migrants, although overstated suspicion about 'bogus' asylum seekers – i.e. asylum seekers who are really in pursuit of employment or other financial gains – is never far from view in the British context (see Zimmermann, 2014, for an exposition of the poverty of the notion of bogus asylum seeking). For the most part in this book I examine the situation of asylum seekers and not economic migrants, although I recognise that there are difficulties and sensitivities in distinguishing between the two.[3]

The British public's attitude towards migrant deaths has been largely insensitive since at least the early 2000s. Occasionally, the magnitude of a disaster or the horrific circumstances that surround it will make the news and provoke a popular, although usually short-lived, sense of guilt, as in the case of the tragic drowning of the toddler Aylan Kurdi, washed up on a Turkish beach in 2015, which prompted a social media outcry and a flurry of grassroots activism, obliging the Prime Minister David Cameron to accept more Syrian refugees to Britain. But most migrant deaths make little impact on public consciousness. UNITED[4] has kept a 'List of Deaths' since 1993, which includes all reported deaths that have occurred as a consequence of European border militarisation, asylum laws, poor accommodation conditions, detention, deportations and carrier sanctions. The fatality count stood at 22,394 by mid-June 2015, although the actual figure is likely to be much higher as a result of the number of unreported deaths (UNITED, 2015). The United National High Commission for Refugees (UNHCR) (2014) reported that 3,419 people lost their lives trying to cross the Mediterranean in 2014 alone, making it the deadliest sea crossing route in the world. Yet because these numbers accrue steadily they have little impact. Until recently, there had been no sustained outcry from the British public against the lethal consequences of the current management of border controls beyond the protestations of a small number of interest groups.

Although this lacklustre attitude might be uncomfortable to acknowledge, it is possible to understand how it originates. Reports of migrant deaths refer to migrant struggles and lives that seem alien to, and distant from, the lives of most citizens in Western developed countries. It is difficult to appreciate their experiences of loss and suffering, especially when the accounts reference far-flung places that are unfamiliar and carry little resonance for the majority of middle-class Westerners. While this should not be taken as an excuse for the persistence of highly securitised border controls that pose a threat to the lives of migrants, it does render intelligible public apathy in the face of the calamities that befall migrants.

The degree of neglect exhibited by the guards, medical personnel and centre managers responsible for Mr Dvorzac at the time of his death, however, goes beyond the more general listlessness of the British public towards migrant deaths. It displays a level of unconcern and a disregard for suffering that is qualitatively distinct from public indifference. Disconcertingly, Mr Dvorzac was well known to the authorities: guards did not 'discover' him in the same way that border control officers came across the migrants in shipping containers. Rather Mr Dvorzac died as a result of neglect by individuals who could see his discomfort, were acquainted with him, and had the power to alleviate his distress. Tragically, other deaths in British detention display similar symptoms. The Institute of Race Relations documents a series of deaths of detainees in British detention between 1989 and 2014, pointing toward the slowness of authorities to react to cries for help, the aggravating role of neglect when medical conditions are already being suffered, misplaced medical records, allegations of poor treatment and assaults by staff, referrals by medical staff that were never followed up, and insufficient care taken to prevent suicides (Athwal, 2014[5]).

It is a gruesome feat to be able to engender, within employees, levels of indifference that allow them to overlook the suffering of subjects right before their eyes. I call this a feat because it must have been achieved despite our tendency to feel weaker empathy for people who are far away from us and stronger empathy for those close to us. The British public's generally lack-lustre response to migrants' suffering can be explained by this tendency: the fact that most migrant struggles occur in settings, countries and situations unfamiliar to most Western citizens, including the ports, docks and vessels that form the backdrop of the deaths in shipping containers and at sea, means that news of migrant deaths seems decidedly removed from their everyday lives. Mr Dvorzac, however, died in full view of the authorities that were supposedly caring and responsible for him and he was not, at the time, attempting to dodge these authorities but was rather relying on them for his welfare. His death, and the deaths of others who have died in similar conditions in detention in the United Kingdom, provides a starting point for my exploration of the relationship between indifference, moral distance and proximity in this book. What interpersonal, institutional and political factors, I ask, are producing levels of indifference that are proving lethal to migrants around the world? And what can anti-border activists do in response to them?

Moral Distance and Encounters

The relation between distance and indifference has been formally conceptualised in terms of 'moral distance'. Moral distance is a concept that enjoys considerable currency among moral philosophers, sociologists and psychologists, and represents a prominent example of geographical language that has been taken up outside the discipline of geography. My intention in adopting it is not to engage in subjective moralising, but to use it to refer to an empirical phenomenon. It refers to the 'distance decay' that moral concerns exhibit, resembling gravity to the extent that people further from us exert a weaker moral claim upon us (Tronto, 1987, citing Hutcheson, 1971; see also Smith, 2000).[6] Put simply, it refers to the human tendency to care more for people close to us than to those far away.

Of course not all distance is the same. Zygmunt Bauman (1989) helps to disentangle various forms of distance and in so doing augments the 'moral distance' argument. In his much-discussed study of the Holocaust[7] he distinguishes the physical from the psychological distancing effect of bureaucratic organisational forms, although both are able to quash 'the moral significance of the act and thereby pre-empt all conflict between personal standards of moral decency and immorality of the social consequences of the act' (Bauman, 1989, p. 25). He also discusses the importance of mediation – that is the density of middlemen and women, or technological devices, that stand between the issuing of an order or the making of a bureaucratic decision and its consequence. Where this density increases, moral estrangement also increases, bringing with it the risk that individuals will be licensed to act immorally in the absence of any clear view of the suffering that their actions may cause. Although Bauman points to different forms of distance though, in essence the moral distance argument involves a consistent claim: that where distance of one sort or another separates individuals, any moral sentiments they might feel for those influenced by their actions are suppressed roughly in proportion to the distance itself.

Consistent with the notion of moral distance, it seems to follow that when distance is overcome this can act as a catalyst to moral concern. In recent years much has been written about 'the encounter'. For the philosopher Emmanuel Levinas (1979, 1981), encounters mean that I[8] come face to face with suffering others[9] such as asylum seekers fleeing persecution, and at this point I become responsible for them and accountable to them, experiencing their bearing of their vulnerability to me as both a plea and a command to respond. It is the face of the suffering other that generates this moral effect. Levinas is careful not to reduce being face to face with someone to merely sighting them: he understands proximity in a specific way that has an ethical rather than an empirical or literal meaning. Nevertheless, he makes clear that there is something morally demanding about being in proximity with someone who is suffering, and authors such as Bauman (1993) and Hamblet (2011) have extrapolated from this observation to make more practical claims about distance, morality and bureaucracy (see also Hamblet, 2003). For Hamblet (2011, p. 717) 'Levinas frames ethics as a problem of distance; the moral challenge is a challenge of geography.' For Bauman (1993, p. 83) '[p]roximity is the realm of intimacy and morality' whereas 'distance is the realm of estrangement and the Law'. Basing his argument on Levinas, Bauman opposes the moral potential of the face to face encounter with impersonal systems of bureaucratic rule that distance officials from subjects.

Border scholars have been largely silent of the topic of moral distance and indifference. In the next chapter I begin by making the case that our understanding of the spatial organisation of borders, border control and border work could be enriched by taking into account their importance. According to this argument the opening of moral distance – that is the phenomenon of *moral distancing* – is an important consequence of the broad shape of recent changes to both 'the state' in general and to modern immigration control systems. The pursuit of efficiency and the smooth operation of systems, the turn to governance, the internationalisation and outsourcing of immigration controls, and the privatisation of large swathes of the business of control, have moral consequences that have been generally overlooked. In the case of British border control, they tend to keep decision makers and asylum seekers apart through various forms of distance and mediate more densely between them, with the effect that the moral check afforded by encounters and 'rights of presence' (Amin, 2002a, p. 972) is extinguished. The ability of 'the sufferer [to] find her way into the direct perceptual range of the moral agent in order to awaken the moral sensibility that will elicit a compassionate response to her suffering' (Hamblet, 2011, p. 717) is seriously undermined by modern border controls.

This keeping-apart makes excluding migrants by force a morally less demanding task. Individual functionaries and managers are not confronted by the worst consequences of their work by having to look their subjects in the eye. By functionaries I mean the frontline personnel who make daily decisions about asylum seekers and who have responsibility for asylum seekers' day-to-day welfare, and by managers I mean the designers and orchestrators of the system of asylum governance that is currently in place.[10] The international obscuring of asylum seekers also ensures that publics in destination countries are insulated from the moral claims of would-be immigrants and the disturbing moral consequences of pre-emptive, remote and forceful border controls.

It would be inaccurate to claim that the restructuring of the state and of border controls in recent years has been explicitly undertaken with this aim of moral distancing in mind. To make this claim would be to credit the managers of state institutions and border controls with more organisational competence than they have ever demonstrated, at least in a British context. It would also, more broadly, risk feeding the 'state-phobia' that Foucault (2008, p. 76) has identified, which has the potential to disseminate a misleading image of an unassailable, monstrous, calculating and coherent state behemoth, possessing 'a sort of generic continuity' (Foucault, 2008, p. 187), that is difficult to resist. Rather, while the drivers of this trend towards moral distancing might occasionally be premeditative and calculative, they are more often mundane and banal, associated with the achievement of immediate targets, the minimisation of costs and the adoption of efficient organisational models and business practices. Moral distancing arises, then, as a result of the dispassionate organisation of practices in accordance with bureaucratic concerns. In this light we might say that moral distancing is an emergent property of a complex system that governs human mobility – a property of the system that is not reducible or traceable to the actions of any individual or parts within it (see Urry, 2007).

Calculated or not, however, separation between decision makers and asylum seekers nonetheless leads to the former becoming detached from the real-world experiences of the latter, to the extent that they are often unable to appreciate the gravity of their own work. This detachment is particularly damaging given that those fleeing the threat of persecution have usually already experienced fearful and traumatic events. The emotional, psychological and economic buffeting that slow, impersonal and detached bureaucratic treatment delivers often acts to compound these difficult experiences.

Enriching Accounts of Moral Distance

Moral distance describes a basic, fundamental consequence of the strengthening and proliferation of national borders in contemporary society and, using this concept, *Nothing Personal?* offers an empirical examination of indifference and immigration control. The primary thrust of the book is to understand the empirically evident indifference to suffering others in Britain's immigration system, and the concept of moral distance provides a useful tool for doing so. Before going further though I want to critically enrich the perspective of moral distance in order to help to formulate a full picture of the indifference towards suffering others generated by border controls and to lay the foundations for the investigation that follows. I will do so in four ways: (i) by addressing the distancing of officials from migrants and not simply vice versa; (ii) by questioning the ethical potential of closeness; (iii) by exploring forms of indifference that are *not* generated by distance; and (iv) by thinking critically about the relationship between indifference and emotions.

We should not assume, firstly, that moral distancing is primarily a matter of distancing subjects on the one hand from publics, managers and functionaries on the other, and not vice versa. The way the restructuring of border control has been discussed by scholars recently, with reference to the export of borders (Clayton, 2010) and the 'push back' of migrants (Bialasiewicz, 2012, p. 856), for example, emphasises *migrants'* experiences of remoteness. But there are other ways in which moral distancing can occur – not simply by alienating A from B, but also B from A. Distance is a relational concept and so it makes sense to consider the experience of distance from the perspective of both parties.

A second important nuance of the moral distance argument is to recognise that literal closeness will not necessarily lead to a morally demanding encounter. It would be easy, but over-simplistic, to assert that where distance is eradicated encounters occur. On the contrary, modern border control systems are also capable of entertaining closeness whilst suspending moral proximity and encounter. It is therefore centrally important, I argue, to think about ways in which encounters are avoided, averted and suspended even when decision makers are close to their subjects. This requires thinking about the different forms of organisational and institutional distance that permit, and often guarantee, moral estrangement at close quarters.

A third necessary elaboration of thinking in terms of moral distance is to recognise that moral distance nurtures only one specific type of indifference. In particular, whereas moral distance operates through the removal of subjects from moral purview, it is possible for indifference to also arise through over-familiarity with suffering others. In making this argument I turn to Simmel (1903/2002) in order to develop a vocabulary around the blasé functionary, whose indifference towards others is of a qualitatively different nature to the indifference that moral distance nurtures. Being alive to the different sources of insensitivity and indifference that combine within complex systems of control is essential to fully understanding them.

A fourth development of the theme of moral distance is to be wary of associating moral distance, and the indifferent, impersonal disposition of the bureaucrat, with a lack of emotion. According to Bauman (1989) bureaucracies tend to produce moral distance through various mechanisms, which allow their functionaries to treat their subjects dispassionately, indifferently and unemotionally. Weber (1948) also associates bureaucracies with emotional coolness. In contradiction of Bauman and Weber, however, the bureaucratic processes in evidence in Britain's asylum system do not rely upon the evacuation of emotion that they set out. Rather, bureaucracy and sensitivity are woven together in subtle and insipid ways in the area of asylum seeker management, which ultimately leads to the strengthening of bureaucratic modes of rule. The management of asylum seekers is able to present a 'softer side' that actively encourages and enrols emotions such as care and empathy among its functionaries and managers and throughout its structure.

This last assertion requires attention not only to the way bureaucracy might co-opt emotion, but also to the way activists might position themselves in relation to the struggles of asylum seekers. In particular, in the closing sections of the book, I consider the implications for progressive border activism of the fact that discourses of care and compassion have been adopted by the systems governing asylum seeker and refugee issues in the United Kingdom. This melding of subjugation and care, repression and compassion, renders any activist attitude towards asylum seekers couched in terms of 'caring-for', 'supporting', 'helping' or 'caring-about' also at risk of co-optation. This brings me ultimately to advocate for activist tactics that are in *solidarity-with* asylum seekers and refugees in the United Kingdom, because it is through this type of language and positioning that activists can ensure that they remain oppositional to, rather than facilitative of or complicit in (however unwittingly), the governance of asylum seekers in the United Kingdom and the passivity with which they are often portrayed.

Asylum Seekers in the United Kingdom

In this book I delve into the working lives of immigration personnel in order to investigate a series of questions. First and most importantly I ask how indifference towards migrants is produced in border control systems. This leads to numerous further questions such as: What are the moral effects of recent changes to border control systems? How are immigration personnel nurtured in such a way as to make them capable of, and willing to, deliver an increasingly exclusionary and brutal system of control? To what extent, and how, are the consequences of their work precluded from them? How are they 'kept apart' from their subjects and through what forms of distance? In exploring these questions *Nothing Personal?* offers a comprehensive study of the relationship between British immigration control, distance and indifference towards suffering others.

Before I can describe my methodology in detail it is necessary to set out the social and political context of the asylum system in the United Kingdom. With this background I can explain how I approached the study of immigration control. In this section I briefly describe recent trends in Britain's asylum system, the media climate surrounding asylum in the United Kingdom, the legal innovations that have impacted upon the asylum sector in recent years, and the recent technical and practical policy initiatives that have come into force.

The United Kingdom is witnessing a sustained intensification in the way systems of governing asylum seekers act to exclude them, govern them through discomfort, criminalise them and expose them to uncertainty and risk (see Vickers, 2012; also Darling, 2011a). In 2002, the United Kingdom received 84,132 applications for asylum. By 2014 this number had fallen to 24,914 representing a 70.4% reduction. In contrast, although the number of asylum applicants to the EU-27 fell from 421,470 in 2002 to just below 200,000 in 2006,[11] numbers subsequently rose to 626,710 in 2014 (a 48.7% increase on 2002 levels) largely due to significant increases in numbers of applicants from Syria, Eritrea, Kosovo, Afghanistan and Ukraine. These changes occurred in the context of many more people fleeing persecution, conflict, generalised violence, and human rights violations globally. In 2002 around 40 million people were forcibly displaced worldwide, but by the end of 2014 nearly 60 million were, constituting levels of displacement that are 'unprecedented in recent history' (UNHCR, 2015: 5).

As a result of Britain's apparent hospitality crisis, the share of forcibly displaced people globally who apply for asylum in the United Kingdom has dropped precipitously. Taking the ratio of the number of asylum claims received by the United Kingdom to the global population of concern to the UNHCR as a crude measure, this ratio fell from 4.1/1000 in 2002 to just 0.5/1000 in 2014.[12] This reduction in asylum claims received by the United Kingdom is the result of the nation's increasingly harsh and exclusionary discourse around asylum migration.

The media climate surrounding asylum seekers in the United Kingdom is central to understanding their treatment. Since at least the early 2000s the popular printed tabloid press (henceforth 'the press') has disseminated a perception that Britain offers generous social security benefits to asylum seekers. The press has also attained notoriety for its heavy-handed, subjective and derogatory treatment of asylum seekers over this period[13] (Mollard, 2001, Leveson 2012). Britain is now routinely perceived as a 'soft-touch' for migrants who supposedly seek out the most attractive reception conditions among European countries. Although asylum seekers' ability to do this has been discredited (Day and White, 2001; Robinson and Segrott, 2002) these concerns endured for over a decade (Kelly, 2012), attaining the status of a full-blown 'invasion complex' (Tyler, 2013, p. 87).

Spurious connections between asylum seekers and a variety of social ills have simultaneously become commonplace. For example, concern has been expressed that asylum seeking is linked to terrorism – 'Bombers are all sponging asylum seekers' the *Daily Express* printed (Daily Express, 2005).[14] Other tabloid sources have exaggerated the cultural mismatch between asylum seekers and British communities with stories that depicted asylum seekers as strange and outlandish. 'Swan Baked... Asylum seekers are stealing and eating swans' *The Sun* reported (The Sun, 2003), whereas the *Daily Star* published the claim that 'Asylum seekers ate our donkeys' (Daily Star, 2005).[15] Others have been concerned that asylum seekers might commit crimes in British host communities: 'Our town's too nice for refugees...they will try to escape, rapists and thieves will terrorise us' the *Daily Express* quoted in a headline (*Daily Express*, 23 March 2002, p. 1) while others are outspoken about supposedly bogus asylum seekers arriving in Britain in order to benefit from the welfare entitlements available to asylum seekers: 'we resent the scroungers, beggars and crooks who are prepared to cross every country in Europe to reach our generous benefits system' *The Sun* has printed (The Sun, 2001).

Although unfounded, concerns that asylum seekers were 'sponging', or terror threats, or culturally mismatched, or represented criminal risks, put pressure on successive governments to control what was quickly conceptualised as the asylum 'problem' and the asylum 'threat' during the 2000s (see Squire, 2009), providing the grounds for greatly toughened policies. The budget allocated to the enforcement of immigration law has increased markedly since the late 1990s, for example. In 1996–7 the Immigration and Nationality Directorate had 5,868 staff and a budget of £218 million. By 2004–5, there were 15,002 staff and the budget had increased to £1.7 billion.[16] This increase in detection and enforcement capacity has been combined with a series of legal innovations designed to make Britain a more inaccessible place internationally and a more hostile place once it has been reached. For example, asylum seekers' access to legal appeals against negative decisions on their claims for asylum has been significantly curtailed via a series of exclusions from access to the appeal system, and cuts and restrictions to legal aid (see Webber, 2012).

Alongside legal measures, a number of technological and practical innovations have also been introduced. In 2005 the government announced a five-year strategy, the key proposals of which included heavy investment in technological capacity, such as large X-ray scanners capable of detecting human stowaways in moving vehicles, electronic fingerprinting, digital scanning of the iris, and the electronic tagging of asylum seekers already in the United Kingdom (Home Office, 2006). The strategy proposed the granting of temporary leave to remain rather than permanent refugee status wherever possible, and fast-tracking of asylum claims[17] so that the time and resources spent on the legal system are reduced. The proposals also introduced the rollout of e-borders, where all international passengers are electronically checked before they reach the United Kingdom, as they enter and as they leave, and a redoubling of efforts to remove unsuccessful asylum applicants in order to achieve parity between the number of those refused and those removed.

The consequences for the asylum seekers who are refused and who might, under different conditions, have been granted asylum are often dire. There are reports that some are tortured and killed upon return to their origin countries, although systematic research into the fatality rate of deportees is sorely lacking in the British context.[18] There are, nevertheless, a series of observed consequences that deported asylum seekers experience. In the case of Afghan deportees, for example, these include 'the impossibility of repaying debts incurred by migration…the shame of failure, and the perceptions of "contamination"' (Schuster and Majidi, 2013, p. 221). For those that remain in the United Kingdom without status, they can expect to endure exploitation (Vickers, 2012), destitution (British Red Cross, 2010), ostracisation and marginalisation among Britain's working-class communities (Hynes, 2009) and defamation in Britain's press (Finney and Simpson, 2009).

Then there are those migrants, like Mr Dvorzac, who lose their liberty in immigration detention facilities as a result of their journeys. When 19-year-old Bereket Yohannes was found hanged in a shower block at Harmondsworth removal centre in January 2006, 61 detainees at the centre issued a catalogue of complaints and indictments of the conditions in removal centres in the United Kingdom. They referred to 'dehumanising and depressing conditions' (Garcia et al., 2006, p. 15), the way in which staff 'make us feel that we are an inconvenience' (Garcia et al., 2006, p. 15), the food that 'would be rejected by some dogs in the United Kingdom' (Garcia et al., 2006, p. 16) and '[t]he way and manner officers disrespect detainees [which] is quite disgusting and very humiliating' (Garcia et al., 2006, p. 16). Nearly ten years later similar issues persisted. A television news investigation aired in 2015 (Channel Four, 2015) included undercover footage of guards at the

Yarl's Wood centre for females showing contempt for detainees, such as by referring to them as 'animals', 'beasties' and 'bitches'.

Regarding disrespect for asylum seekers among border control officials, Louise Perrett, a former employee of UKBA, blew the whistle on the tactics used by staff at a major centre for processing asylum seekers' claims in the United Kingdom in 2009. She identified practices of mistreatment, trickery, humiliation, generalised hostility, indifference and rudeness among staff. According to her account, when claims were complicated she was advised simply to refuse them, and when immigration staff granted 'too many' claims then a humiliating 'grant monkey' (a soft toy) was placed on the desk of the culprit (Taylor and Muir, 2010).

The British government also routinely suffers embarrassing public relations disasters that have occurred because individuals working within immigration control either make mistakes or step out of line. Scandals have included the mistaken release of hundreds of convicted criminal migrants who should[19] have been considered for deportation under British law (BBC, 2006b); the employment of asylum seekers, who were not supposed to undertake paid employment according to British law, in Immigration and Nationality Directorate (IND) offices (BBC, 2006c); and evidence that senior officers have tried to exchange immigration status for sexual favours (Doward and Townsend, 2006).

Substandard treatment of migrants by staff and poor public relations have been linked to deep-seated cultural deficiencies at the heart of the government institutions that oversee border control. Commentators have detected widespread denial that asylum seekers might be positing legitimate claims. Denial refers to 'an advance decision to avoid situations in which … facts might reveal themselves' (Cohen, 2001, p. 23). The UNHCR, for example, has diagnosed a 'refusal mindset' among decision makers (UNHCR, 2005, p. 17). Asylum claims have been refused without properly considering the facts of individual cases or the country of origin information that is made available to decision makers, and by using speculative arguments and citing a small number of peripheral inconsistencies as grounds to dismiss entire applications (Amnesty International, 2004; Amnesty International and Still Human Still Here, 2013). As a result fully 25% of initial decisions are eventually overturned on appeal, indicating the wastefulness of the initial decision making process even on its own terms. New members of staff are plunged into this system with little training and either have to acculturate rapidly or face the psychological and professional consequences of swimming against the tide.

Approaching Immigration Control: Spaces and Settings

Researching the way border control decision makers, including frontline officers, elite managers and contracted agents, relate to migrants is no easy matter because access is often highly constrained, especially around secure sites such as Immigration Removal Centres. This is due, in large part, to anxiety among managers and gatekeepers that research will either not be in their interest, will lead to some sort of public embarrassment or that it could compromise the security of such centres. Many functionaries, for example, are contractually forbidden from discussing their work because doing so might involve security breaches. A related methodological challenge is obtaining a clear overall view of the system of controls. Functionaries tend to be positioned in specific roles and often do not have a view of the entire system. Managers, on the other hand, can be less knowledgeable about the everyday, on the ground, happenings at particular sites of border work.

Another challenge concerns the relentless 'policy churn' meaning the 'endless stream of new initiatives' (Hess, 1998, p. 52) that characterises British immigration control. In terms of legislation, a major new piece of legislation has been introduced in the United Kingdom every couple of years over the past 20 years, which often significantly rewrites immigration rules, causing confusion for migrants and support groups and providing a challenging research environment.[20] In comparing the immigration control systems of the United States and United Kingdom, Bohmer and Shuman (2008) point out that whereas the US system has been slow to adapt to changes in international relations, the UK system has been, 'if anything, too quick to change' as a result of the fact that 'rules and laws, unlike in the US, are not subject to constitutional oversight' (Bohmer and Shuman, 2008, p. 22). Maiman (2005, p. 244) is similarly disconcerted by 'the British government's...unchallengeable capacity to make, unmake, and remake its own rules'. This has been reflected not only legislatively, but also in the frequent creation and disbanding of institutions that oversee border control in the United Kingdom. In 2007 the Immigration and Nationality Directorate was replaced by the Border and Immigration Agency, which was replaced in 2008 by the UK Border Agency (UKBA), which was itself abolished in 2013 in order to return the work of immigration control to the Home Office.

I approach these challenges using three general principles that have underpinned a programme of research that began in 2003 (methodological details relating to timescale of research activities, access, sampling, analysis and ethical considerations are provided in the Appendix). Firstly, I have employed a range of qualitative methods on the premise that different methodologies give different insights into the complex phenomena under study. *Nothing Personal?* therefore draws upon interviews, focus groups and ethnographic work as well as document and policy analysis in an attempt to form a nuanced picture of the objects of the research. I draw most frequently upon interviews, which have been conducted at various points through the research period. Interviews were generally recorded unless the interviewee explicitly requested that I did not use a voice recorder, which was sometimes the case among anxious immigration personnel (I discuss anxiety among immigration personnel in greater detail in Chapter Six).

The interview was then transcribed in full if it had been recorded, or else written out as extensively as possible on the basis of scratch notes taken during and immediately after the unrecorded interviews. Transcripts were then sometimes shared with the interviewee for approval, correction or elaboration. They were then coded according to a set of research themes that I had distilled in advance from existing academic literature and my own research questions, and that guided my approach to the varied empirical material that multiple methodologies generate. Focus groups were held in London in 2012 and brought together activists, charity workers and volunteers for a series of compelling conversations (see Tyler *et al.*, 2014). The ethnographies were conducted in 2013 and 2014 by my researchers Drs Melanie Griffiths and Andrew Burridge, who spent considerable time observing asylum appeal procedures in the first tier immigration and asylum tribunals in various tribunals around the UK.

Secondly, I have avoided confining the analysis to any single site of immigration control. Although there are various excellent studies that focus upon individual sites of border work such as detention centres or the interview process, *Nothing Personal?* provides an overview of the British immigration system by examining a series of relevant settings. These span key sites in the execution of the different stages of an asylum application, including the main site of initial claims processing in the United Kingdom at Lunar House in Croydon, London, the location of back-office work relating to asylum claims for welfare support in Portishead near Bristol, and Campsfield House Immigration Removal Centre near Oxford, where individuals are detained under immigration powers, ostensibly pending their removal from the United Kingdom. These sites differ not only according to their formal function in asylum claim determination and immigration enforcement processes, but also according to their political sensitivity and media profile. Lunar House was the target of sustained media scrutiny through much of the 2000s, for example, and as such finding willing interviewees there and gaining access to the site was more challenging even than accessing Campsfield Detention Centre. By contrast I occasionally found staff employed in other areas and sites of immigration control surprisingly willing and eager to participate in my research, sometimes in order to vent their frustration about their working conditions. More broadly, by taking an approach that spanned multiple research sites, the book is able to identify general, system-level patterns in the way officials are governed and the way that indifference is nurtured.

Thirdly, given the challenges of gaining a clear overview of the system, a variety of groups have participated in the study. So although I do draw on research with frontline decision makers, contracted security staff, police officers, back-office employees and elite immigration managers, I also draw upon evidence from migrants themselves who have experienced indifference and insensitivity first hand, as well as activists, charity workers and community leaders. One noticeable phenomenon in this respect is the degree to which some individuals occupy more than one subject position. For example, I have interviewed police officers who are also activists, refugees who are also government workers, and solicitors and government workers who are involved in multiple initiatives that are often very different and sometimes in tension. Often these different subject positions would only come to light part way through interviews, but they serve to highlight the difficulty of firmly categorising individuals, and often gave me pause for thought about my own preconceptions.

Plan of the Book

Nothing Personal? proceeds over seven further chapters. In Chapter Two I set out the case for taking account of the morally distancing consequences of border work. I outline the moral potential of proximity from a variety of disciplinary viewpoints, and examine how recent rounds of state and border rescaling and restructuring have made proximity between decision makers and asylum seekers less likely. The chapter consequently calls for a rereading of modern state rescaling through the lens of its interpersonal effects and makes the case that moral distancing, and the indifference to suffering that it promotes, is a primary consequence of recent changes in border control practices.

Identifying the moral distancing effect that changes to the bureaucratic management of borders has in the international context is important, but needs to be approached carefully. This account of moral distancing does not help to account for Mr Dvorzac's death, for example. Chapters Three, Four, Five and Six therefore draw on varied empirical material to develop critical reflections on some of the assumptions of the broad picture presented in Chapter Two. These chapters offer important embellishments to the account of moral distance, especially with regard to the ability of systems of control to nurture moral indifference of decision makers towards subjects even when they come close to each other, when they are in contact for considerable periods of time and when they feel significant emotional attachment to each other. Together they highlight the different forms of distance that keep decision makers and asylum seekers apart, and the different forms of indifference operating throughout border control work.

Chapter Three examines the importance of thinking about moral distance from the perspective of both partners in a relationship of distance. The chapter examines the remarkable extraction of asylum decision makers from the environments in which asylum seekers were present through the 2000s in the United Kingdom, as part of a drive to regionalise and modernise asylum support and decision-making systems. Offices and employees were located well away from the urban concentrations of migrants thereby insulating them from contact with their subjects. Contracted agencies were positioned between them and they were set into competition with each other over abstract metrics that gave no clue as to the human gravity of the activities they undertook. In this way distance between functionaries and subjects was opened not by excluding subjects but by removing functionaries from contact with migrants.

Chapter Four examines the situation in which physical distance has been overcome by considering the cases of asylum interviews and asylum appeals.[21] These are contact events that are both legally required and that represent the most effective way to exchange the sort of complex information that it is necessary to exchange in the determination of individual asylum cases. Here functionaries and asylum applicants come close to each other, but what is striking about these meetings is how rarely they entail morally demanding encounters. Somehow, the ethical epiphany that Levinas describes in proximity is suspended. The chapter draws on the psychological literature on contact to identify the intricate ways in which indifference is nurtured, and encounters suspended and averted, even at close quarters.

In the case of immigration detention, which is the subject of Chapter Five, the indifference of functionaries towards their subjects is sustained even during prolonged contact. Any notion that physical proximity alone might provoke moral sentiments is consequently thrown into question. In detention, overstimulation of the empathetic instincts of personnel is commonplace, caused by their overexposure to harrowing accounts of trauma and prompting them to adopt tactics of psychological avoidance as a form of self-care. The incessant churning of detainees exacerbates this exposure, whereas their trivialisation, infantilisation and repeatedly asserted strangeness make aloofness towards them easier still. Avoidance thus morphs from a spatiotemporal phenomenon to a psychological one. Perversely, it is the very closeness of staff to detainees that achieves this effect (Simmel, 1903/2002, p. 14).

Chapter Six refutes an important assumption that beleaguers theorists of indifference and insensitivity: that indifference towards others can be associated with a lack of emotion. From Bauman and Simmel to Glover and Weber, the Chapter begins by setting out evidence of this widespread conjecture. Yet there are at least two emotions that functionaries commonly experience that serve to actually facilitate rather than frustrate the development of indifference. The first is anxiety, which nagged almost every functionary I came across or heard about during the course of my fieldwork. Without anxiety – over discipline from managers or embarrassment in the press – many more functionaries might have the imaginative courage to overcome their own insensitivity. And the second, perhaps more disconcertingly still, is care. The ability of immoral systems to interweave care and indifference in increasingly complex ways, allowing functionaries to morally question their involvement *and find themselves blameless*, signals a higher level of sophistication in the development of insensitivity than these theorists can accommodate.

Chapter Seven turns to activist attempts to counteract indifference and insensitivity among immigration personnel. Drawing on the experiences and tactics of a subset of migrant support organizations, the chapter describes mobilizations that seek specifically to nurture compassion among functionaries and decision makers. Such activities aim to directly confront the impersonality and indifference of bureaucratic border control by repersonalising elements of the system – an approach that, I argue, entails a series of risks. Nurturing compassion requires closeness to institutional centres of control, and the spectre of co-optation is never far from view in these situations. The pursuit of compassion among functionaries, which entails metaphoric and sometimes literal pleading with them, also signals a capitulation to the structure of the system that bestows these functionaries with power and authority in the first place. Most fundamentally though, given that compassion and sensitivity are perfectly compatible with brutal systems of control (as I demonstrate in Chapter Six), making the nurturing of compassion among functionaries an activist objective risks strengthening the system itself. The chapter provides some illustrations of this and, although it lists a series of mitigating considerations and extenuating circumstances that might render compassion-seeking less risky and more worthwhile, it concludes by questioning the conditions that have reduced some activists to pity-seekers and setting out the demanding conditions under which activism in pursuit of compassion is desirable.

The conclusion provides a summary of the argument of the previous chapters and synthesises the key insights that the book develops regarding the generation of indifference towards asylum seekers amongst border officials. Beginning from this empirical starting point, the book sheds light on the various forms of indifference operating in British immigration control, the opportunities and limitations of thinking about changes to immigration control systems in terms of moral distance, the techniques by which encounters are suspended or averted even in situations of face to face and sustained contact, and the co-optation of softer and gentler discourses in the brutal business of border control management.

Notes

1   Although most people in British immigration detention have sought asylum in the UK at some point (The Migration Observatory at the University of Oxford, 2015a), Mr Dvorzac himself was not seeking asylum, he had just become confused when asked by border officials where he was travelling to, resulting in his detention. His treatment is indicative, though, of the sort of treatment that it is possible to receive in immigration detention.

2   The tragedy was the subject of a Hong Kong, Cantonese language film, *Stowaway* (2001), shot in Fuzhou, Vietnam, Moscow, Ukraine and England.

3   It is worth noting that the term 'asylum seeker' has become associated with a range of negative connotations and tends to paste over different national experiences in an unhelpful way. Alternative terms are therefore arguably more appropriate, such as 'sanctuary seeker', 'refugee' or simply 'migrant', the latter of which rejects the notion that distinctions need to be made between migrants on the basis of their reasons for migrating. Although I retain the term asylum seeker in this book because it was in such wide usage among both my participants and the legal and policy sources I draw from, the deficiencies with the term 'asylum seeker' should consequently be borne in mind throughout.

4   The European network against nationalism, racism, fascism and in support of migrants and refugees.

5   http://www.irr.org.uk/news/deaths-in-immigration-detention-1989-2014/

6   Throughout the book I follow Proctor (1999) in understanding morality to be concerned with 'the *normative* sphere of human existence and practice' (Proctor, 1999, p. 3, italics in original) as opposed to ethics, which refers to 'systematic intellectual reflection on morality in general, or specific moral concerns in particular' (Proctor, 1999, p. 3).

7   I accept that the Holocaust was an historical event of unparalleled atrocity and magnitude in the recent history of Western developed countries and I am not suggesting that immigration detention in Western countries is comparable to the Nazi death camps.

8   Levinas sometimes writes in the first person, which has the effect of increasing the impact of his prose.

9   I use this term 'Other' to describe those considered different and unfamiliar. The term is a general one and need not imply suffering or neediness, although for the most part in this book I use the term to refer to Others who are also in some form of need.

10 I use the term 'decision makers' to refer to managers and functionaries collectively.

11 Non-EU-27 applicants only.

12 Figures quoted in this paragraph and the previous one are taken from UNHCR (2002), The Migration Observatory at the University of Oxford (2015b), Eurostat (2015) and UNHCR (2015).

13 In fact the *Daily Mail*, Britain's best-selling tabloid newspaper, has exhibited a staunchly anti-immigration stance for over 70 years. In 1938 it published the following: '"The way stateless Jews from Germany are pouring in from every port of this country is becoming an outrage...." In these words, Mr Herbert Metcalfe, the Old Street magistrate, yesterday referred to the number of aliens entering the country through the "back door" – a problem to which the *Daily Mail* has repeatedly pointed' (*Daily Mail*, 20 August 1938; see Karpf, 2002, for a fuller discussion).

14 The article about the 7 July bombers was inaccurate – the identity of the bombers was unknown when the story was written and neither of the men mentioned in the story was an asylum seeker anyway.

15 Both of these stories were simply untrue. They are made up, but were front page news. *The Sun* published the following clarification over five months

later without forewarning on page 41 of its newspaper (see Medic, 2004): 'A report in *The Sun* on the 4th July about the disappearance of swans in southern England stated that asylum seekers were responsible for poaching them. While numerous members of the public alleged that the swans were being killed and eaten by people they believed to be Eastern European, nobody has been arrested in relation to these offences and we accept that it is not therefore possible to conclude yet whether or not the suspects were indeed asylum seekers'. And the Leveson Inquiry into media practices found that the story about the donkeys was 'total speculation' and that the police had 'no idea what had happened to the donkeys' (Leveson, 2012, Vol. 2, Sect. 8.47).

16 The budget remains at around this level despite the number of asylum seekers requiring support reducing significantly since the mid-2000s, implying that more resources have been directed towards detection, deterrence and prevention mechanisms. Data on expenditure on border control and enforcement taken from http://www.theguardian.com/news/datablog/2012/dec/04/government-spending-department-2011-12, http://www.publications.parliament.uk/pa/cm200506/cmselect/cmhaff/775/775i.pdf and Back *et al.* (2005).

17 Fast-tracking proved particularly controversial. The Detained Fast Track (DFT) was a system designed to expedite the determination of asylum claims. From the early 2000s to mid-2015 the United Kingdom operated a fast-track asylum process according to which individuals could be taken straight from the port of entry to detention to have their claim decided quickly. Even if a refused claim was appealed, denied and appealed again to either the High Court or the Court of Appeal, the entire process was scheduled to take just 21–22 days (although in reality it often took longer). Serious questions over the impartiality and thoroughness of these procedures were raised (Asylum Aid, 2013). The UNHCR, for example, recorded 'concerns regarding the quality of decisions made within the DFT, including the concern that the speed of the DFT process may hinder the ability…to produce quality decisions' (UNHCR, 2008, p. 24). It noted the use of standard wording to refuse claims without engaging with the specific circumstances of particular cases as well as frequently inaccurate application of key refugee law concepts in the DFT setting. The DFT was found to be unlawful by a High Court judge in June 2015 and was suspended soon afterwards, a decision that was upheld by the Court of Appeal in July 2015.

18 This said see www.lifeafterdeportation.com for an attempt to collate deportees' experiences.

19 The British government aims to remove foreign national offenders as quickly as possible to their home countries, ostensibly to protect the public, to reduce costs and to free up spaces in prison. It should be noted, however, that removal often occurs at the end of a prison sentence, thereby constituting double punishment for a single offence. It is also very easy for foreign nationals to infringe complex immigration laws and become branded as criminals whilst posing no threat to the public.

20 Major pieces of legislation introduced in the past 20 years include: the Asylum and Immigration Act 1996; the Special Immigration Appeals Commission Act 1997; the Immigration and Asylum Act 1999; the Nationality, Immigration and Asylum Act 2002; the Asylum and Immigration (Treatment of Claimants, etc.)

Act 2004; the Immigration, Asylum and Nationality Act 2006; the UK Borders Act 2007; the Borders, Citizenship and Immigration Act 2009; and the Immigration Act 2014 (Great Britain 1996; 1997; 1999; 2002; 2004; 2006; 2007; 2009; 2014).

21 By asylum interviews I mean both the screening interviews and substantive interviews that form parts of the determination process in the British system. I explain the distinction between these two types of interview in Chapter Four. Asylum appeals are legal events, held in tribunals, at which immigrants put their case to an immigration judge.

Chapter Two
Moral Distance and Bureaucracy

The greatest evil is not now done in those sordid 'dens of crime' that Dickens loved to paint. It is not done even in concentration camps and labour camps. In those we see its final result. But it is conceived and ordered (moved, seconded, carried, and minuted) in clean, carpeted, warmed and well-lighted offices, by quiet men with white collars and cut fingernails and smooth-shaven cheeks who do not need to raise their voices.

C.S. Lewis, *The Screwtape Letters* [preface]

On 19 June 1945 a Soviet woman named Nataliya[1] knelt on the floor of the Foreign Office in London at the feet of a civil servant and, through desperate tears, pleaded for her life, the life of her infant and the life of her husband. The bewildered Foreign Office official, Mr Brimelow, was responsible for overseeing the deportation of Soviets back to the Soviet Union who had been captured as prisoners of war during the final phases of World War II by British and allied forces. In the Soviet Union they were considered traitors. Nataliya's Soviet husband, Ivan, had been captured by Nazi forces in 1942 and forced to fight for the Germans against the Soviets. Despite his reluctance to fight, the terms of the Yalta agreement between Britain and the Soviet Union dictated that he and his wife must be returned, notwithstanding the fact that it was common knowledge that Soviet 'traitors' would almost certainly be executed or sent to work in a labour camp when they arrived. Many Soviets had already been deported in this way, but Nataliya's

Nothing Personal?: Geographies of Governing and Activism in the British Asylum System,
First Edition. Nick Gill.

case was contested because she had a 6-month-old child who had been born in England. She had also been fortunate because Ethel Christie, an activist and agitator on behalf of the Soviets, had taken up her case and arranged the meeting at the Foreign Office. In his book, which details how the order to force desperate Soviets onto trains carrying them to their deaths took a wrenching toll on British soldiers in the 1940s, Nicholas Bethell outlines the extraordinary effect that Nataliya's emotional plea had upon Mr Brimelow and his colleagues. 'Foreign Office officials are not used to such scenes ... Nataliya had touched their hearts during her distressing visit to the Foreign Office, and for the first time they were allowing considerations of humanity and morality to enter the argument' (Bethel, 1974, p. 50). Mr Brimelow petitioned his superiors in the Home Office to make an exception in Nataliya and Ivan's cases and, despite strong political pressure to comply with the Yalta agreement in order to ensure that all British prisoners of war were similarly returned to Britain by the Soviet Union, the Foreign Office yielded and the family built a life for themselves in England. As Bethell notes, 'their story was one of very few with a happy ending' (Bethel, 1974, p. 52).

What is it about proximity that can provoke empathy or compassion and embolden bureaucrats to bend the rules? And what is it about distance that is capable of suspending moral engagement and nurturing indifference to the plight of others? The notion that we care less for people who are far away from us, either literally or owing to some form of social or cultural distance, is deeply ingrained in Western thought. Aristotle, for example, observed that 'the nearness of the terrible makes men pity ... sufferings are pitiable when they appear close at hand, while those that are past or future, ten thousand years backwards or forwards, either do not excite pity at all or only in a less degree' (Aristotle, 1926 trans., *Rhetoric* II.8, 1386a). Similarly, Hume famously stated that 'The breaking of a mirror gives us more concern when at home, than the burning of a house, when abroad, and some hundred leagues distant' (Hume, 1739/1896, 2.3.7). And Adam Smith, in reflecting upon selfishness and human reason, opined that, should the whole of China be 'suddenly swallowed up by an earthquake' (Smith, 1790, III.I.46), a man of humanity in the West might reflect sorrowfully on the situation and might reason about its causes, but 'when all this fine philosophy was over, when all these humane sentiments had been once fairly expressed, he would pursue his business or his pleasure, take his repose or his diversion, with the same ease and tranquility, as if no such accident had happened' (Smith, 1790, III.I.46).

Glover (1977) conceptualises the natural human tendency to compartmentalise our perceived ethical responsibilities and to limit them to proximate issues in terms of moral distance. 'Moral distancing', he writes, 'can be plausibly explained as defence mechanisms that serve a valuable purpose. There is a psychological need for clear categories ... anyone with a steady imaginative grasp of the avoidable evil in the world would suffer some kind of psychological collapse' (Glover, 1977, p. 290). Tuan concurs, arguing that 'people need ... mental boundaries between self and other so as not to be overwhelmed' (Tuan, 1999, p. 115). From this perspective, part of the explanation for Mr Brimelow's petition on behalf of Nataliya lies with the fact that Nataliya punctured the boundary between Mr Brimelow's 'compartments' during her emotional visit to the Foreign Office, and passed from being a faceless member of a categorical group to being a particular and very real person right before his eyes, who therefore had a moral claim upon him through her very presence.

In this chapter I explore the issue of moral distance as it relates to administering the quintessential form of modern rule – bureaucracy – and border controls in particular. Philosophers, sociologists and psychologists have, in different ways, recognised the influence that proximity can have over moral decision making, a consequence of which is the potentially disruptive nature of moral encounters to bureaucratic procedures. When the arrangement of border control mechanisms reduces proximity and the likelihood of encounters between decision makers and subjects it therefore affects the morality of these mechanisms. This is exactly what recent rescaling and changes in the governance arrangements of states and border controls has achieved.

I establish the moral potential of proximity in the first section, and in the second section, drawing on Zygmunt Bauman's work, I explore the notion that there is a tendency within administrative bureaucracies to preclude moral proximity between bureaucrats and subjects. In the third section I interpret a variety of recent developments of the modern state in terms of this tendency. I highlight the moral implications of the state's 'upward', 'downward' and 'outward' rescaling (Brenner, 2004) in terms of the opening of moral distance and the avoidance of disruptive encounters between decision makers and subjects, and in the penultimate section I highlight the same implication with respect to the specific area of state activity concerned with immigration control. By setting out this broad account I call attention to the interpersonal implications of contemporary state rescaling and border control practices, and prepare the ground for a series of critical reflections on this account that draw on my empirical evidence in the chapters that follow.

The Status of Proximity in Moral Theory and Practice

Peter Singer (1972) urges us to resist the natural spatial moral myopia that facilitates moral distance. For Singer physical proximity and distance should not figure in calculations of moral responsibility, and we should respond just as sensitively to the suffering of distant others as to those who are close to us. Yet most philosophers disagree with Singer owing to the hugely impractical demands that morality without some form of distance decay would make on individuals. Indeed, in some cases the overlooking of distance may itself have an immoral effect. In *Bleak House* Dickens humorously illustrates this point through his comedic figure Mrs Jellyby, an English social philanthropist so wrapped up in writing letters in support of orphanages in Africa that she abandons her own children and ignores them even when they hurt themselves and cry out for her help (Dickens, 1852–1853/1993). Friedman (1991, p. 818) makes the point more seriously: 'Hardly any moral philosopher … would deny that we are entitled to favour our loved ones. Some would say, even more strongly, that we should favour them, that it is not simply a moral "option"'. If we recognise the necessity of partialism[2] though, a challenge immediately arises because it is necessary to choose how to restrict our obligations to others in some way. Moral theorists have searched for rules and principles to help guide individuals through this dilemma (for an accessible review, see Sandel, 2009). Should we direct our attention and our energies towards those in worst need, for example? Or to those with whom we share some sort of relationship? Or to those who are least able to care for themselves and those around them?[3]

One answer is to give those most proximate to us greater moral weight. Some moral theorists, for instance, have drawn on Christian philosophy, and specifically the story of the Good Samaritan, to make the point that there is something morally demanding about needs that confront us *now*, with immediacy and presence (Waldron, 2003; see also Garber, 2004). Jesus's parable about the Good Samaritan cautions against discriminatory forms of partiality on the basis of ethnic ties or pre-existing relationships (see Box 2.1 for the parable). By the end of the parable both the priest and the Levite are condemned because they did not stop to help the suffering man. By contrast, the Good Samaritan is condoned because he recognises his moral duty to stop and help despite the fact that he shares no particular affinity with the sufferer. We are not told that the Samaritan is a universal do-gooder nor are we led to believe that he is someone who has devoted his life to caring about distant others or pursuing universal human rights in an abstract sense. He is, in this sense, partial. But what he does respond to is what he happens upon. Jesus uses the principles of chance encounters and proximity to determine who should show compassion and generosity, and when.

In Jesus's parable it is the corporality of the encounter that does work in establishing moral momentum: the *seeing*, the up-close, 'in your face'-ness in the words of Jeremy Waldron (2003, p. 350). Levinas has similarly underscored the moral significance of the face to face encounter. In his book *Totality and Infinity*, Levinas (1979) develops a conception of the self that, in its presocial state, is self-centred and concerned only with self-gratification in its relations with the world. 'In the ordinary world of everyday life things and people are there for me, the I, to use, consume, enjoy and thereby to

Box 2.1   The Good Samaritan

And, behold, a certain lawyer (Nomikos) stood up, and tempted him, saying, Master, what shall I do to inherit eternal life? he said unto him, What is written in the law? how readest thou? And he answering said, Thou shalt love the Lord thy God with all thy heart, and with all thy soul, and with all thy strength, and with all thy mind; and thy neighbour as thyself. And he said unto him, Thou hast answered right: this do, and thou shalt live. But he, willing to justify himself, said unto Jesus, And who is my neighbour? And Jesus answering said, A certain man went down from Jerusalem to Jericho, and fell among thieves, which stripped him of his raiment, and wounded him, and departed, leaving him half dead. And by chance there came down a certain priest that way: and when he saw him, he passed by on the other side. And likewise a Levite, when he was at the place, came and looked on him, and passed by on the other side. But a certain Samaritan, as he journeyed, came where he was: and when he saw him, he had compassion on him, [a]nd went to him, and bound up his wounds, pouring in oil and wine, and set him on his own beast, and brought him to an inn, and took care of him. And on the morrow when he departed, he took out two pence, and gave them to the host, and said unto him, Take care of him; and whatsoever thou spendest more, when I come again, I will repay thee. Which now of these three, thinkest thou, was neighbour unto him that fell among the thieves? And he said, He that shewed mercy on him. Then said Jesus unto him, Go, and do thou likewise.

Luke 10: 27–37, King James version

become nourished' (Morgan, 2011, p. 63). Yet the one-to-one encounter with another challenges this self-centredness. Encountering the other forces the self, at a fundamental level, to take account of others.

This encounter, for Levinas, is the activation of an 'originary impulse' (Dikeç *et al.*, 2009, p. 6) that precedes any form of calculation of the costs and benefits of interaction. It occurs before knowing the individual concerned, and, in the context of immigration control, before 'check[ing] to see if each others' papers are in order' (Dikeç *et al.*, 2009, p. 6). This is not to say that it is based on similarity with the other. In fact, Levinas was at pains to point out the rupture that an encounter causes, which arises as a result of the profound *difference* between self and other that proximity lays bare. Barnett (2005) has consequently underscored the asymmetry of the self's relation to the other in a face to face encounter, referring to the 'traumatic exposure to alterity' (Barnett, 2005, p. 9) that is involved in this 'most intense and singular experience of difference' (Barnett, 2005, p. 10).

Often, however, the other is overlooked in everyday life so that this type of encounter is either avoided or prevented. 'What is occluded, hidden, forgotten in our ordinary lives is ... this presence of the other's face to me – and my responsibility to and for this person' (Barnett, 2005, p. 59). This is why the face to face encounter, and proximity, are so important to Levinas (1979, 1981). They act as sharp reminders to ensure that the self is 'put in question by the alterity of the other' resulting in a specific form of 'exposedness' (Levinas, 1981, p. 75) of the self to the plea and demand of the other that Levinas calls 'nudity' (Levinas, 1981, p. 89). Exposure to another's need and suffering, for Levinas, prohibits my turning away. Aptly for our purposes, Levinas uses the metaphor of the homeless refugee to illustrate the obligatory nature of this sort of face to face encounter:

> He has no other place, is not autochthonous, is uprooted, without a country, not an inhabitant, exposed to the cold and heat of the seasons. To be reduced to having recourse to me is the homelessness or the strangeness of the neighbour. It is incumbent on me.
>
> Levinas, 1981, p. 91

Hence, 'my nonindifference to the neighbour' (Levinas, 1981, p. 91) is mandatory wherever the face of the other breaches intentional everyday consciousness and confronts me with its destitution. Breaches like these render me 'obedient as though to an order addressed to me. Such an order throws a "seed of folly" into the universality of the ego' (Levinas, 1981, p. 91).

Anything less than a face to face encounter is not sufficient to this task. 'There is no way to correctly and precisely sever that appeal and command from the way she looks at me; that appeal and demand is *in* the look, so to speak' (Morgan, 2011, p. 66, italics in the original). This is because, above all, the face to face encounter is 'utterly particular' (Morgan, 2011, p. 59): it does not reside in theoretical or abstract relations. Rather, 'When one person faces another, she does so as a dependent person ... It is a targeted dependence, my dependency upon you' (Morgan, 2011, p. 71). The 'face' for Levinas is thus more than just the presence or appearance of the other. It is the other's bearing of their own vulnerability and destitution to me that carries with it a command not necessarily to reciprocate, help or even recognise the other, but entails an experience of difference that subjects the self (Barnett, 2005). The face, consisting of exposure to the suffering of another, acts as a solemn summons of the self to respond that cuts through everyday rhythms and routines. In his later work Levinas (1981) also writes about the disruptiveness of the face, likening an encounter with the face of the other to an assault on our complacency that is capable of disturbing our distracted musings and awakening our morality.

It is important to note some disagreement over how to interpret the encounter in Levinas's thought. For some it is an ethical event, and does not necessarily refer to a situation of spatial contiguity. The 'reception of the other' Dikeç *et al.* (2009, p. 8) warn, 'is not a simple spatial opening or inclusiveness'. Barnett (2005) concludes that 'it might not be best to take Levinas' work too literally as an account of practical conduct' (Barnett, 2005, p. 8). Others, however, interpret Levinas's thought more literally, and in particular with reference to the practical implications it might have. Given that spatial metaphors and concepts like proximity and the 'face to face' pervade his writing, it is possible to understand why, for example, Wendy Hamblet (2011, pp. 717–8) declares that '[f]or Levinas geographical immediacy is the only factor that has a claim upon conscience' (see also Hamblet, 2003). Bauman also draws associations between Levinas's concepts of proximity and the face to face on the one hand, and physical closeness on the other (Bauman, 1989).

If it is the case that proximity has a moral or ethical dimension, this position is certainly in keeping with our psychological impulses. Two traditions of thought in psychology, on obedience and on contact, underscore the way in which literal proximity figures not only in our moral reasoning but also in our moral practice. Regarding obedience, Stanley Milgram's (1974/2005) experiments involved asking subjects to administer electric shocks of rising intensity to a pupil who they thought was another recruit to the experiment but was actually an actor. The subject was told that they would have to administer a stronger shock for each mistake that the pupil made in a word-pairing exercise in order to allow scientists to determine how learning was affected by punishments. In fact the frequency of 'mistakes' was preordained (and no shocks were actually given). Of the 40 participants in the original study, 26 proceeded to the highest voltage level (450 volts). Sometimes they expressed concern, but some verbal prodding from the experimenter was sufficient to convince them that this was indeed in the interests of science and that the experiment required them to continue. Milgram then varied the experiment by 'bringing the victim closer' (Milgram, 1974/2005, p. 34). First, while the victim stayed in an adjacent room, voice-protests were audible to the subject. Further iterations of the experiment brought the victim closer still; they entered the same room as the subject and in the closest iteration the subject had to literally force the hand of the victim onto a metal shock plate. The proportion of subjects that defied the experimenter and refused to take part in the experiment any longer on the basis of their concerns for the 'victim' rose from 35% for the initial experiment, to 37.5% in the voice-feedback situation, 60% in the proximity situation and 70% for touch-proximity. As Milgram notes, the most disturbing thing about these findings are the proportions of people who persisted right to the end of the experiment in each case. We can conclude, though, that our moral sentiments are at least partly conditioned by a certain form of distance. The closer we are to someone the more likely we are to object to their cruel treatment.

Regarding contact, Gordon Allport's (1954) seminal study suggested that contact between different groups may lead to reduced prejudice if certain conditions were met. This is because cognitive bias can be reduced through proximity to, and increased contact with, people whom you consider different from yourself. Although psychologists since have stressed the strenuousness of these conditions, Allport's hypothesis that contact can reduce prejudice has endured (Torre *et al.*, 2008; Askins and Pain, 2011. We return to Allport's contact hypothesis in Chapter Four).

A range of figures in moral psychology and philosophy, then, have recognised the significance of proximity to moral matters. It is interesting, however, that many geographers have been more cautious in approaching the relationship between proximity and morality. Dikeç *et al.* (2009) observe, for instance, how easy it is to be close to others in today's cities without ever morally encountering them: to be 'eye to eye but worlds apart' (Dikeç *et al.*, 2009, p. 2) from the lives of subjugated others in our midst. They therefore outline the need 'to engage more carefully with the "proximities" that prompt acts of hospitality and inhospitality' (Dikeç *et al.*, 2009, p. 1). Valentine (2008), and others, take an analogous approach by interrogating the conditions under which meetings and co-presence constitute 'meaningful' contact – a concept I explore in greater detail in Chapter Four. In a similar vein, Barnett (2005, p. 11) challenges us to disrupt the 'homologies often drawn between spatial proximity, partiality and care on the one side, and spatial distance, impartiality and justice on the other'.

One way in which this challenge has been taken up is in the work of Doreen Massey (2005) and Ash Amin, who have charted the relational networks that link local places such as cities to far-flung sites. '[S]o extensive have the city's connections become' write Amin and Thrift (2002, p. 26) 'as a result of the growth of fast communications, global flows, and linkage into national and international institutional life, that the city needs theorization as a site of local-global connectivity, not a place of meaningful proximate links'. Their relational approaches have emphasised the varying geographical reach of interpersonal connections, releasing such connections from literal locality and co-presence. Indeed, they are critical of the 'nostalgia' that they detect in much geographical work that is critical of the distances that supposedly produce alienation, dysfunction and anomie, and instead argue that various kinds of 'distanciated communities' (Amin and Thrift, 2002, p. 41) come into view '[o]nce we move away from the notions of face-to-face or heavily localised interactions as the only kind of authority' (Amin and Thrift, 2002, p. 43). Such an awareness of the connections between localities and far-flung others is all the more important in the internet age, in which 'there is no logical reason to suppose that moral boundaries should coincide with the boundaries of our everyday community: not least because these later boundaries are themselves not closed, but rather are defined in part by an increasing set of exchanges with distant strangers' (Corbridge, 1993, p. 463). Even face to face communication need not involve literal proximity, Amin and Thrift (2002) argue, since 'the face is increasingly mobile' (Amin and Thrift, 2002, p. 38) via screens and fast, mass communications.

There are also acknowledgements of the continuing moral potential of literal physical proximity and encounter among geographers, however. For David Smith (1998), care is a local and situated affair. The increasing academic prominence of the feminist ethic of care is symptomatic of what he sees as a 'resurgent partiality' (Smith, 1998, p. 27) in the face of the impossibility of the sort of universalism prescribed by Peter Singer. 'Care ethics raises caring, nurturing, and the maintenance of interpersonal relationships to the status of foundational moral importance', its proponents argue (Friedman, 1993, p. 147). Understanding care as a specific, embodied relationship with someone who is known to the carer, Smith draws associations between the feminist ethic of care and communitarianism (Smith, 2000). '[B]oth imply partiality' he argues, 'favouring members of our own family, group or community. What is more, they both appear to rely heavily on the proximity of the persons concerned, to make their morality work' (Smith, 2000, p. 83). Popke (2006, p. 507) reiterates this observation, reasoning that if 'relations of care are affective, embodied and relational then an ethics arising out of this would seem to be necessarily partial and situational, holding only for those with whom we have some immediate contact and familiarity'. This view is implicitly more sceptical than Amin and Thrift (2002) about the ability of new telecommunications technologies to extend the ambit of face to face, intimate interaction.

What is more, even among proponents of relational approaches to understanding spatial relations, there is also evidence of the persistent influence of the local and the importance of the face to face that runs alongside the interconnections between the global and the local that they highlight. In light of the increasing heterogeneity and diversity of many of the world's cities, alongside the tensions that this can produce, Amin has argued that tolerance at the level of everyday, prosaic negotiations of difference within local micro-publics can be an antidote to ethnic tension (Amin, 2002a). Whereas theorists of citizenship, race and ethnicity have tended to focus upon the national and international levels when discussing rights and obligations, Amin argues that local negotiations of difference mean more to the people that experience them than abstract theorising. This is why segregation in the city is so damaging: because, from the beginning, 'the very possibility of everyday contact with difference is cut out' (Amin, 2002a, p. 969). Although everyday contact with difference is not a panacea, it can offer the opportunity for different and supposedly opposed groups to 'step out of their daily environments into other spaces acting as sites of "banal transgression"' (Amin, 2002b, p. 12). This has implications for the design of cities, which can be calibrated to ensure that the paths of different groups intersect more frequently. The potential of an everyday encounter is that it can challenge the preconceived, abstract and categorical understanding of another group through interaction with *this particular* person in *this* place and time who happens to be different from me. It is 'the *practice* of negotiating diversity and difference' (Amin, 2002a, p. 971) that an encounter offers. In parallel with Levinas's notion of the command that the face issues, Amin urges us towards a vocabulary of the 'rights of presence' (Amin, 2002a, p. 972) that urban encounters invoke.

For Massey also, the global city throws up exciting possibilities with respect to the 'happenstance juxtaposition of previously unrelated trajectories' (Massey, 2005, p. 94) such as 'the business of walking around the corner and bumping into alterity' (Massey, 2005, p. 94).[4] She argues that the modern diversity of urban space is distinctively chaotic, and that this chaos presents both a risk and a chance for things to change. Like Amin, this 'throwntogetherness' of the city (Massey, 2005, p. 151), as she puts it, necessitates 'the *practising* of space ... where negotiation is forced upon us' (Massey, 2005, p. 154). Although she is careful, like Amin, to emphasise that chance urban encounters are not, in themselves, likely to solve the problem of ethnic tension in modern cities on their own, since any politics of place also involves a 'wider ... politics of connectivity ... of openness and closure' (Massey, 2005, p. 181), her approach invokes excitement over the potential of the sheer being-together-ness of the city, which can 'enable "something new" to happen' (Massey, 2005, p. 94). 'One answer to the question of crisis', writes Aitken (2010, p. 58) in discussing her work, 'is a fundamental and emotive acknowledgement of diverse spaces of encounter in a chaotic throwntogetherness'.

Keeping People Apart

If we accept the moral potential of literal proximity, what implications does this have for the way we understand bureaucratic forms of organization? It seems clear that the bureaucratic mode of organisation, which requires dispassion, impersonalism and detachment, can only be expected to accommodate a certain measure of proximity before these traits are eroded. After all, from the perspective of a bureaucratic administration, some of the side-effects of proximity sound particularly irksome. Why would system managers embrace the 'dislocation' and 'surprise' that Aitken (2010, p. 64) associates with throwntogetherness? Why would they risk disobedience or the generation of fondness for subjects that the psychologists like Milgram posit as a by-product of proximity when a bureaucracy requires dispassion-ateness? Why would they expose their employees to additional obligations in the light of the 'rights of presence' (Amin, 2002a, p. 972) that proximate subjects possess? Why would they seek out the conditions in which something new and unpredictable might happen when a bureaucracy relies upon smooth and predictable flows of information and resources? Rather, would they not baulk at the thought of Levinas's (1981) 'commands' being issued to their own functionaries during a face to face encounter, lest these demands run counter to the bureaucracy's own instructions to its bureaucrats? Encounters of this kind evidently represent a variety of risks of disrupting administrative procedures. The last thing bureaucracies need to have to manage is the sort of spontaneous, instinctual reactions to situations that Mr Brimelow displayed when exposed to Nataliya. Rather than pursuing the potential of Massey's throwntogetherness in the modern city then, bureaucracies are far more likely to display an aversion to proximity precisely because of its disorderly and unsettling character (Levinas, 1981). Where decision makers and subjects are kept apart this represents an expedient release from the unwelcome disruption that their proximity might bring.

The question that then becomes important concerns how moral distance has been manipulated by institutional arrangements: there is a need to develop a richer understanding of the moral grammar of bureaucratic institutions including border controls. How, for example, have the 'compartments' that Glover (1977) and Tuan (1999) discuss been expanded or condensed? What institutional factors influence whether or not an individual includes a particular plea within their moral frame of reference? The sociologist Max Weber identified bureaucracy as one of the defining features of modern life, and associated the eschewing of personal standards of ethics with the bureaucratic form. Industrialization, he argued, has led to a fixation with the rational planning of modern life, aided by a separation of 'the office' and the attendant figure of the official from the home. As part of this transformation, Weber noted a separation of the means and ends of bureaucratic systems, which are concerned with the pursuit of efficient, rational criteria, from those of moral and ethical systems of judgement. Within a bureaucracy, 'a technical orientation to means and ends always rules out decision making in terms of ethical standards' (Morrison, 2006, p. 379). Instead, the objectives of the bureaucratic institution, and the conditions that allow those objectives to be pursued, are of paramount importance.

> Precision, speed, unambiguity, knowledge of files, continuity, discretion, unity, strict subordination, reduction of friction and of material and personal costs – these are raised to the optimum point in the strictly bureaucratic administration
>
> Weber, 1948, p. 214

A consequence of the primacy of the objectives of the institution over individual mores, therefore, is that personal ethical judgements are evacuated from a proper bureaucratic system. Compared to other types of decision making, Weber argued that bureaucracy was technically superior precisely because it achieved this evacuation. 'Administration by notables and collegiate bodies', explains Morrison (2006, p. 381) in reviewing Weber's work, 'is always less efficient because their interests inevitably conflict and bring about compromises between views. This creates delays which slow down progress and make decision making less precise and less reliable.' In order to avoid differences of opinion, then, bureaucracies are governed by norms of impersonality. Employees act according to their roles of office rather than in terms of their personal ties, and officials treat people only as cases rather than as individuals. As Weber writes:

> The honour of the civil servant is vested in his [sic] ability to execute conscientiously the order of superior authorities, exactly as if the order agreed with his own conviction. This holds even if the order seems wrong to him... [D]enial of the authority of private conscience, become[s] now the highest moral virtue
> Weber, 1948, p. 95

This is not to say that frontline staff always capitulate to this bureaucratic code of honour. One of the major practical challenges for bureaucracies is how to inculcate the ethic of subordination of self-interest to organisational objectives among bureaucrats. Despite thorough training, this has historically proven troublesome. In his widely referenced study, Michael Lipsky (1980) distinguishes between system managers and what he calls 'street-level bureaucrats' who have decidedly different objectives. They tend to 'have an interest in minimizing the danger and discomforts of the job and maximising income and personal gratification' (Lipsky, 1980, p. 18). 'These priorities', Lipsky continued, 'are of interest to management for the most part only as they relate to productivity and effectiveness' (Lipsky, 1980, p. 18). Weber's description of bureaucracies therefore rarely manifests itself fully because frontline workers have minds of their own. Moreover, although it is certainly not their priority (that being their own interests), street-level bureaucrats enjoy 'considerable discretion' (Lipsky, 1980, p. 23) and can 'intervene on behalf of clients' (Lipsky, 1980, p. 23) according to 'personal standards of whether or not someone is deserving' (Lipsky, 1980, p. 23).

Nevertheless, for Lipsky, the 'intrinsically conflictual' (Lipsky, 1980, p. 25) nature of bureaucracies arises primarily as a result of the struggle between workers' and bureaucratic interests. The parameters of this struggle are set largely by the design of institutional incentives and disincentives, the perceived legitimacy of management's demands upon workers, and the organisation and room for manoeuvre of street level bureaucrats themselves, rather than by any consideration of moral responsibility towards subjects or clients. Lipsky locates personal moral standards of fairness beneath the demands of self-interest and the demands of management in a modern bureaucracy. For both Weber and Lipsky, then, the nature of the quintessential modern organisational form of the bureaucracy seeks and tends to extend moral detachment. The volume of concerns that fall within the moral purview of the individual decision maker is systematically reduced through bureaucracy.

It could be argued that this is precisely what is required from a bureaucracy. Du Gay (2000), for instance, understands Weber's description of a detached and impersonal bureaucrat is an ideal type rather than a critique. Yet it is also clear that the impersonality of bureaucracy can sometimes become pathological. Zygmunt Bauman (1989) has explored the dehumanisation and moral invisibility of the Jews and other persecuted groups to the perpetrators of atrocities in Germany during the Holocaust, emphasising the confluence of three factors that facilitated such spine-chilling results. First he identifies distance itself. Once defined as different and undesirable, much of the activity of the Third Reich consisted in separating, cordoning and concentrating the Jews into neighbourhoods, then ghettos, then camps. This distancing, for Bauman, achieved what he calls *estrangement* as the 'offending category' is 'removed beyond the territory occupied by the group it offends' (Bauman, 1989, p. 65). In its turn this estrangement achieves a sinister result. Rather than having to struggle morally over the right way to treat an undesirable group, distance alleviates the necessity to morally scrutinise actions because they are conceived of abstractly and their consequences are obscured from view.

The second factor he emphasises is mediation. Drawing on John Lachs's (1981) work, Bauman argues that a central aspect of distanciation is the degree to which one's own actions are separated from their final consequences by either middlemen or women, or through technologies that distort a clear view of the causal link from one to the other. In the case of the separation of actions from consequences via middlemen or women, where there are long chains of command from the ordering of an action to its actualisation, those at the top of the chain have nothing but an abstract notion of the sorts of results their orders and authorisation are likely to achieve, whereas those at the bottom of the chain can tell themselves that they are simply following orders. Taken together, this means that almost no one takes responsibility for the organisation as a whole. Hand in hand with a long chain of command goes an increasingly fine functional division of labour. Where individuals are engaged merely in small, technical and repetitive tasks that bear very little resemblance to the overall outcome (perhaps they are making the canisters to carry the gas, or producing the axles that support the trains), they are more able to assuage any passing guilt they may have by juxtaposing their seemingly inconsequential actions to the larger whole.

In the case of the mediating effect of technology, Bauman equates modern technologies that are able to reduce humans to 'graphs, datasets [and] printouts' (Bauman, 1989, p. 116) to 'moral sleeping pills' (Bauman, 1989, p. 26).

Technology's ability to distance cause and effect reaches its zenith in contemporary weaponry that kills across vast distances. But the same result occurs when decision makers who hold the lives of others in their hands meet them merely as 'statistics' (Bauman, 1989, p. 99) such as 'variables, percentages, processes and so on' (Bauman, 1989, p. 116), which succeed in 'obliterating the humanity of [their] human object' (Bauman, 1989, p. 115).[5]

In such situations, the decision maker, freed from any direct encounter with their subject that is not heavily abstracted, is thus made available to pursue technical objectives and measure the results 'without passing any judgement, and certainly not moral ones' (Bauman, 1989, p. 99). 'Thanks to rapidly advancing new information technology' Bauman (1989, p. 115–6) concludes, 'psychological distance grows unstoppably and on an unprecedented pace'. Key to the distanciation of the Jews was the removal of specific, individual Jews from German society so that, by degrees, real Jews that Germans knew were replaced by an abstract category of Jew, what Bauman calls the 'metaphysical Jew' (Bauman, 1989, p. 189). In the end, 'the Jew was only a "museum-piece"… something one had to journey far to see' (Bauman, 1989, pp. 189–90).

Whilst it is important not to view all technologies as equivalent, some geographers have explored similar ground in terms of the ability of technology to 'distance' the consequences of contemporary violence from its causes. Louise Amoore discusses the modern use of pre-emptive algorithmic surveillance technologies to identify suspicious movements and activity before a crime is committed, as well as X-ray scanners, which have become a common sight at border control crossing points. These technologies can be understood to 'mine the body for certainties' (Amoore and Hall, 2009) by dissecting the human person into its constituent parts, embodying an insipid type of violence that is 'concealed in the glossy techno-science' (Amoore, 2009) of border control. Derek Gregory (2013), in discussing the use of military drones by the US army, offers further insights into the distancing capabilities of technology. He notes a series of features of drone warfare that bear an unsettling resemblance to the effects that Bauman and others associate with moral atrocities, including 'the dispersion of responsibility across the network' (Gregory, 2013) and the importance of drone operators' abilities to psychologically 'set aside' (Gregory, 2013) the consequences of their actions and 'partition' (Gregory, 2013) the moral implications of their work.[6]

A third factor Bauman outlines is the ability of the Third Reich to equate rationality and immorality. One of the most sinister achievements of the Third Reich for Bauman was its ability to harness '[i]ndividual rationality in the service of collective destruction' (Bauman, 1989, p. 135). He notes incredulously 'how few men with guns were needed to murder millions' (Bauman, 1989, p. 202) and traces this feat to the orchestration of rational action in the pursuit of destructive objectives. He gives examples of the meagre, but vital, benefits accruing to Jews who occupied social positions that greased the wheels of the Jewish extermination itself. He concludes that under the horrific conditions of the Holocaust 'rationality of self-preservation was revealed as the enemy of moral duty' (Bauman, 1989, p. 143).

Overall, Bauman alerts us to a variety of tendencies, which he sees as inherent to bureaucracies, that can dehumanise subjects to the point of obscuring their moral worth. What we learn from Weber and Bauman is that bureaucratic modern administrations seek neither immoral nor moral bureaucrats, but amoral ones, driven by technical considerations that systematically evacuate personal ethical considerations from the business of carrying out bureaucratic work. In order to achieve efficiency of rule, governments render and evaluate actions as *adiaphoric*, that is 'neither good nor evil, measurable against technical (purpose-orientated or procedural) but not moral values' (Bauman, 1989, p. 215; see also Bauman and Donskis, 2013). For Bauman, this is a tendency of all bureaucracies, not just dictatorships. Although there are checks and balances in liberal society, these contain and limit but do not eradicate the tendencies of bureaucracy.

Rereading the Modern State in Terms of Moral Distance

We must be alive to the specific interpretations of key thinkers that Bauman employs. He uses Weber to paint a dystopian picture of bureaucracy, which may not have been his intention and may not be accurate given the variety of uses to which bureaucracy can be put (Du Gay, 2000). He also interprets Levinas more literally than some scholars would permit (Barnett, 2005; Dikeç et al., 2009), which allows him to place an almost naïve faith in physical proximity and co-presence to provoke morally demanding encounters. In the following chapters I will mount a critique of the assumptions that Bauman employs by drawing on empirical evidence, but before doing so I want to use Bauman's perspective to explore the modern organisation of border controls. It transpires that parts of his analysis hold significant resonance with the contemporary organisation of border practices.

Geographers and anthropologists emphasise how disunified, chaotic and incoherent 'the state' is in practice, implying that a high degree of critical reflection is required in relation to how the concept of the state is deployed (Trouillot, 2001; Painter, 2006; Sharma and Gupta, 2006). Taking an anthropological approach, authors have demonstrated its fragility (Gupta, 2006), the extent to which it depends upon people for its enactment (Jones, 2007) and its improvised and provisional nature (Jeffrey, 2013). Their arguments constitute an antidote to visions of a monstrous, methodological behemoth that coldly and coherently suppresses and controls populations at least in functioning liberal democracies (Nietzsche, 1892/1961).

It is for this reason that we need to approach the concept of 'the state' itself cautiously. For the purposes of understanding border control, I prefer Judith Allen's view that the concept of the state is 'too aggregative, too unitary and too unspecific to be of much use in addressing the disaggregated, diverse and specific (or local) sites that must be of most pressing concern' (Allen, 1990, p. 22). My preference is to frame the discussion of British border control primarily around the tendencies inherent to bureaucratic administration. Framing the discussion in this way holds a series of advantages over discussing 'the state'. First, the fact that bureaucratic organisation is commonly to be found in both public and private sector settings allows the concept to accommodate the privatisation and contracting out that typifies contemporary border control practices (discussed below). Second, because bureaucracy has *inherent* tendencies towards depersonalisation, as Weber and Bauman have demonstrated, there is no particular need to accuse 'the state' of premeditated or calculative malice, which – while tempting when writing about migrants' experiences – is often very difficult to substantiate without slipping into the realm of conspiracy theories. It is the nature rather than the design of bureaucracy that leads to the treatment of individuals as units and specimens (arguably, this is ultimately more disturbing because there is no obvious target for resistance). And third, I am drawn to thinking in terms of bureaucracy because it describes both a relationship between ruled and ruler(s) and an institution, affording it considerable agility with which to transgress settled distinctions between structure and agency in much social scientific thinking about the state. In questioning the usefulness of the concept of the state, Allen urges us to attend to 'other more significant categories and processes' (Allen, 1990, p. 34) such as 'bureaucratic culture' (Allen, 1990, p. 22) that may not be lumped neatly into the specific list of organisations and institutions that the liberal and Marxist aggregation of the state refers to, and I concur with these sentiments.[7]

Nevertheless, in this section I engage more closely with the concept of the state than I do in the rest of the book in order to assess a set of ideas that have been developed in relation to it. This requires the adoption of a different vocabulary, around governance and state rescaling, and a set of different geographical metaphors to those of distance and proximity, in the form of 'upward', 'downward' and 'outward' reorganisation. The challenges of taking on these different spatial vocabularies, however, are worth the gains in terms of bringing thinking on moral distance, which is an interpersonal phenomenon, to bear on a set of debates that usually overlooks the interpersonal dimension.

It might be argued that the sort of moral distance that Weber and Bauman describe is dependent upon an outmoded form of decision making – a particular form of bureaucracy that is top-down and hierarchical. My point in this section though is that it is by no means clear that the recent shift from government to governance, which a set of state theorists have identified, should be accompanied by a reduction in moral distance, and in fact there are various reasons to suspect that it might actually aid the opening of moral distance. State theorists tell us that the state has changed in nature over the last 40 years. Jessop (2002) characterises the change in terms of a shift from a Keynesian national-scale state that is concerned with guaranteeing the welfare of its citizens to a state more concerned with getting ahead in the international, long-run race to secure competitiveness through innovation. This sort of state places greater emphasis on the incentivisation than on the welfare of its workforce. Accompanying this shift has been a 'dramatic intensification in societal complexity' (Jessop, 2002, p. 229) that has given rise to the 'resort to heterarchy ... most evident in the explosion of references to networking ... and in the growing interest in negotiation, multiagency cooperation, partnership, stakeholding and so on' (Jessop, 2002, p. 229).

The state is not necessarily weakened under these new arrangements. '[N]ational state institutions continue to play key roles in formulating, implementing, coordinating, and supervising ... policy initiatives, even as the primacy of the national scale of political-economic life is decentred' Brenner (2004, p. 3) observes. Thus, 'the state ... can indirectly and imperfectly steer networks' (Rhodes, 1996, p. 660). '[T]he ideal of the "social state"' Rose (1999, p. 142) explains,

> gives way to that of the "enabling state". The state is no longer to be required to answer all of society's needs for order, security, health and productivity. Individuals, firms, organizations, localities, schools, parents, hospitals, housing estates must take on themselves – as "partners" – a portion of the responsibility for their own well-being.
>
> Nikolas Rose, 1999, p. 142

Under governance arrangements then, the problem facing the state alters to one of how most effectively to nurture the conditions under which agreements and solutions can be reached through self-organisation. The core state function becomes one of steering and conducting rather than one of carrying out. The state busies itself with setting the meta-coordinates of action and designing the architecture for negotiation and engagement between differently positioned partners in governing. This new, meta-governmental role of the state has 'extensive scope' (Jessop, 2002, pp. 52–3) including actions to 'stabilize the cognitive and normative expectations of actors' (Jessop, 2002, p. 230), 'promot[e] a common "world-view"' (Jessop, 2002, p. 230) and engage in 'boundary-spanning roles and functions, creating linkage devices, sponsoring new organizations, identifying appropriate lead organisations to coordinate other partners, designing institutions and generating visions to facilitate self-organization in different fields' (Jessop, 2002, p. 242). Far from obsolescence, Jessop argues, through these roles and functions '[s]tates play a major and increasing role' (Jessop, 2002, p. 242).

The way these mutations of the state influence the relationship between state managers, frontline bureaucrats, non-state stakeholders and subjects is key to understanding the moral grammar of contemporary states. At the most fundamental level, the introduction of a meta-level of state action, focused upon the steering of self-organising systems 'at a distance' (Rose, 1999, p. 120), is only likely to alienate system managers still further from the eventual consequences of their actions. If anything the 'meta-position' that the state assumes under governance introduces a more distant, more highly abstracted, more adiaphorised view through which to set the parameters of negotiation and exchange than under even a top-down system. Under a top-down system at least state managers are connected to the everyday work of carrying out state projects through a line of communication that relays commands and notices relatively frequently. Under governance no such line exists – only sporadic reappraisal of the performance of autonomous self-governing communities of stakeholders.

In terms of the specifically spatial implications of the reorientation of states towards meta-governmental functions, Brenner (2004, p. 67) writes that '[c]ontemporary state institutions are being significantly re-scaled at once upwards, downwards and outwards to create qualitatively new, polymorphic, plurilateral institutional geographies that no longer overlap evenly with one another, converge upon a single, dominant geographical scale or constitute a single, nested organizational hierarchy'. It might be argued that aspects of this 'hollowing out' (Rhodes, 1994) could reduce moral estrangement and bring decision makers closer to their subjects. This case is strongest with regard to the 'downward' transfer of powers. One could posit that this reduces the distance between the site of decisions and the site of activity, therefore closing the moral distance that Bauman identified by cutting out middlemen and women and thereby reducing mediation. 'This downscaling of regulatory tasks should not be viewed as a contraction or abdication of national state power, however', Brenner (2004, p. 62) warns, 'for it has frequently served as a centrally orchestrated strategy'. Typically, local authorities have a limited degree of devolved discretion over how to achieve this or that end, but very little discretion over the ends themselves (Rodríguez-Pose and Gill, 2003). At worst, devolution offers national governments the opportunity to divest themselves of troublesome duties by enrolling local agents as allies in dispatching them. Hence programmes of public service end up being conceived at the national level but implemented by local government agents, consequently actually stretching lines of command and increasing the degree of mediation between cause and effect that Bauman warns against. It is therefore not at all clear that the trend towards devolution promises closure of moral distance: in fact there are reasons to suspect the opposite.

The internationalisation of state power – Brenner's 'upward' shift – looks even less likely to reduce moral estrangement between decision makers and subjects. If meta-governance can be associated with increased functional distance between the steering activities of state elites and the consequences of this influence on the ground, internationalisation achieves more literal physical distanciation between supranational state elites and the sites at which the consequences of their decisions are experienced. As the European Union expands, for example, its core concentrates and insulates discretionary power from an increasingly far-flung and culturally diverse hinterland. Although closeness on its own by no means ensures moral action (as we shall see in later chapters), unfamiliarity and cultural distance certainly pave the way for adiaphorism.

And when it comes to the 'outward' transfer of powers the geographical rescaling of the state under conditions of governance begins to look like a master class in the reorganisation of state institutional spaces so as to create desensitised, rationalistic decision-making environments. Where possible, large parts of the state were privatised as part of the shift to governance in most Western developed economies. Typically this did not involve total independence from the state, however. Either the state was the main purchaser of services from the newly formed private organisations, in which case it was able to dictate the terms of production via contracts that had to be won from the state itself. Through this development states retained the ability to pursue their own agendas, but distanced themselves from the messy business of implementation. Or alternatively, the state retained a role in regulating the new private organisations, often via newly formed or empowered regulatory bodies that concerned themselves with designing the parameters within which private organisations were at liberty to operate (Peck and Tickell, 2002).

Where privatisation was not possible, one particular development bound up with governing at a distance has been the establishment of entrepreneurial ways of relating between bureaucrats, departments, offices and quasi-state stakeholders as new forms of government have evolved (Rose, 1999). As part of this evolution the cultivation of a bureaucratic ethic of self-subordination to the authority of the institution, which was always so problematic, is no longer necessary. It is replaced by an ethos of business within the state itself. According to this ethos 'the focus is upon accountability, explicit standards and measures of performance, emphasis on outputs ... desegregation of functions into corporatized units operating with their own budgets and trading with one another, contracts and competition' (Rose, 1999, p. 150). The key features of this new way of governing the state include 'contracts, targets, performance measures, monitoring and audit' (Rose, 1999, p. 151). The non-inclusion of moral criteria into these targets licenses bureaucrats and quasi-state partners to also exclude such considerations and place them outside their frame of reference.

It could be argued that the increased distance of state elites from the activities of partners under governance creates some room for manoeuvre 'on the ground'. With more responsibility and freedom, local governance partners have more leeway to act on the basis of moral concerns. But note that the state retains the right to 'recentralize control if the operations and/or results of networks do not fulfil the expectations of state managers' (Jessop, 2002, p. 237). Specifically 'states ... reserve to themselves the right to open, close, juggle and rearticulate governance from the viewpoint not only of its technical functions but also from the viewpoint of partisan and overall political advantage' (Jessop, 2002, p. 239). With the constant potential of this sort of rearticulation in mind, over time the modern state *disciplines* its 'partners' in governing to autonomously pursue the objectives that it values by combining the threat of withdrawal with rewarding performance that is commensurate with meta-governmental priorities.

Hence, if under top-down bureaucracy it was not considered good practice, or professional, to introduce personal considerations into the management of public affairs, under the conditions of governance partners both within and outside the state will find that it simply *does not pay* to be moral. So whereas once it did not particularly cost individual bureaucrats to act morally, under modern conditions individual interests and amorality are aligned, making it rational for decision makers to exclude moral concerns on the basis of their own self-interest. Contracts will be lost, budgets slashed, bonuses forfeited, deadlines and targets missed, and reputations eroded until it will become obvious to all of the governing partners who have been given 'freedom' and 'autonomy' to pursue a particular governmental objective that stoically maintaining sensitive, moral ways of acting is not the most appropriate 'solution' and there are more efficient means on offer. An audit will be undertaken in order to identify the behaviour that does not contribute towards, or even actively undermines, the success of the community of stakeholders in pursuing their objectives. Cost-saving initiatives and performance-enhancing reforms will be recommended that will squeeze the last traces of humanity out of their activities. The truth about the shift from government to governance is that it equates adiaphoric action with the self-interest of bureaucrats and other partners in governing.

Although Jessop admits that he 'ignore[s] issues of interpersonal relations' (Jessop, 2002, p. 217), this omission is a significant one. The effect of the spatial rescaling of the state upwards, downwards and outwards in exempting moral considerations from the relations between decision makers and subjects should not be overlooked. Such rescaling has profound psychological consequences for bureaucrats and other enforcers in terms of estranging them from the subjects of their decisions. If, as Rose has argued, an important dilemma for modern, governmental states is 'how [are] ... bureaucrats and civil servants to be governed?' (Rose, 1999, p. 149), the spatial rescaling of the state constitutes an important part of the answer. It draws upon the simple principle of keeping the subjects of state power out of the sight, and hence out of the minds, of its purveyors.

Moral Distance and Immigration Controls

The evolution of developed countries' immigration controls over the past three decades epitomises the principle of keeping decision makers and subjects apart. Mirroring Brenner's (2004) characterisation of the rescaling of the modern state in general terms, the eviction of national-level responsibilities for asylum seekers and other irregular migrants has been described as a coordinated 'remote control' (Guiraudon, 2003) strategy that renders migrants stateless by geographical design (Mountz, 2010). The 'push-back' of responsibility for border control (Bialasiewicz, 2012, p. 856) involves remote detention both within and outside sovereign territory, innovative new forms of interdiction on airplanes and in airports, and pre-emptive border control measures that seek to deter and contain migrants either en route or at source (Mountz, 2010). As with Brenner's analysis of the state in general, we can discern upward, downward and outward tendencies in the rescaling of migration control[8] (Lavenex, 2006).

The upward rescaling of border control has seen coordinated efforts to hold migrants in legally ambiguous offshore or otherwise remote sites before ever reaching sovereign territory to claim protection. Would-be European immigrants are held increasingly frequently in offshore black holes of indecipherable legal status, from which they have no right to continue their journey or to return home (Vaughan-Williams, 2009; Mountz, 2010). Such international cooperation on migration control has been underway at the European level for at least the last three decades, driven by a desire to avoid the legal, social and financial costs of dealing with in-country asylum claims and housing, incarcerating and deporting claimants (Anderson and Den Boer, 1994; Guild, 2000). This began with a process of consultation and information sharing through intergovernmental negotiating forums, international police cooperation, shared databases and the determination of multilateral criteria for entry and 'burden sharing' (Guiraudon, 2000; Lavenex, 2001). Formal intergovernmental partnership began with only a subset of European countries in 1985, but by 1992 the European Community (EC) was making policy for the region, and by 1999 European tools such as directives and regulations were ratified for use in the area of border control (Koslowski, 2006). Today, the European Union's (EU's) common border policies, including the Common European Asylum Policy, have given rise to 'a new ... landscape that has been built at the external borders of the EU that consists of waiting zones, camps, new fences, and new biometric methods of patrol' (van Houtum, 2010, pp. 958–9). A formidable information-sharing and rapid-response security taskforce, FRONTEX, patrols this new landscape, effectively fielding irregular migrants from the legal and moral purview of individual member countries.

As part of this up-scaling of control, the introduction of 'safe third country' rules to the European Union through the 1992 London Resolutions meant that asylum applicants lost the right to choose their country of asylum (Good, 2007). This concept was a milestone in the insulation of core, powerful countries from the messy business of contact with asylum-seeking populations. It meant that asylum claimants who had passed through supposedly safe countries on their way to countries such as the United Kingdom, Germany and France could be returned to them automatically without being considered for asylum in these core countries. This innovation subsequently became European policy, as well as a popular tool globally (Mountz, 2010), meaning that the fringes of Europe were to act as a buffer, protecting the core countries from proximity to asylum seekers. Cooperation with source countries to reduce the number of asylum seekers who leave and expedite the return of those who have left has also been a recurrent feature of the up-scaling of control, formalised through so-called 'readmission treaties'. Between 2004 and 2008 the EU dedicated €250 million to funding readmission agreements with developing countries (Lavenex, 2006). Their clear logic is to eschew, evict and eject asylum seekers as quickly as possible and as far as possible from their intended destinations.

In this vein various European countries have explored the possibility of the wholesale outsourcing of the processing of their asylum applications to centres in northern Africa (Schuster, 2005; see also the discussion in Nethery et al., 2013, of Australia's measures to 'export' immigration detention into Indonesia). Before Gaddafi's ousting, for example, Italy had a well-developed partnership with Libya, which included a 'readmission agreement, training for Libyan police officers and border guards, and Italian-funded detention and repatriation programmes for irregular migrants in Libya' (Andrijasevic, 2010, p. 150). This partnership had the knock-on effect of causing Libya to strengthen its own borders with Niger, in a development that Rigo likened to a transnational corridor of expulsion (Rigo, 2007, cited in Bialasiewicz, 2012, p. 855). Although the details of Italy's relationship with the new Libya are still unclear at the time of writing, Italy's desire to continue to contribute airplanes and other military equipment to guarantee border security, as well as to train border police in Libya, is unmistakable (ANSAmed, 2013).

Such 'upward' relinquishing of responsibilities for asylum prevents would-be asylum seekers from ever making a claim. The requirement, procurement and verification of visas, a set of practices Guild refers to as 'the border abroad' (Guild, 2002, p. 89) similarly results in the enrolment of foreign agencies into the project of preventing migration at source. Visa impositions tend to follow an increase in asylum applications from a country; for example, in the case of Britain, Sri Lanka in 1985; India, Bangladesh, Ghana, Nigeria and Pakistan soon afterwards; Turkey in 1989; Sierra Leone and Ivory Coast in 1994; and Columbia in 1997. The visa system mobilises distant bureaucrats in the operation of functionally specific tasks located at the administrative offices, ports and airports of sending countries (Weber and Bowling, 2002; Neumayer, 2006). At the European level, the development of a long blacklist of countries requiring visas offers an expeditious way to 'grant (or deny) admission before leaving a country and ... control when someone enters and leaves the EU' (van Houtum, 2010, p. 963).

This system frees its functionaries from moral considerations through a fine functional differentiation of roles and tasks that looks disturbingly similar to that which Bauman describes. The enforcement of these pre-emptive regimes produces 'an advance guard of immigration staff' (Gibney and Hansen, 2003, p. 8) including Airport Liaison Officers and Immigration Control Officers, with little or no view of the immigration system as a whole, who are set to work in the pursuit of tightly defined objectives of checking, verification, collection, surveillance and policing in dispersed settings.

In this vein, Mountz (2013) characterises the phenomenon of detaining asylum-seeking populations in remote, offshore, extraterritorial locations as part of states' tendencies to intercept asylum seekers and bar them from core areas. Within and beyond sovereign territory, remote, often rural or ex-military holding facilities are used to disperse and conceal migrants from 'family members, friends, co-workers, resources, and potential advocates' (Mountz, 2013, p. 91). In particular, islands such as Lampedusa, Guam, and Christmas Island 'offer extreme forms of dispersal, keeping potential asylum seekers at a distance from sovereign territory where they could make an asylum claim' (Mountz, 2013, p. 98), often well beyond the purview of immigration decision makers such as judges and system managers.

For its part, the 'downward' transfer of responsibility for immigration control from national to sub-national actors is best described not as devolution, but as delegation. In general, sub-national legislation has increased the responsibilities of local authorities in checking the legal status of immigrants, providing welfare support for asylum seekers and other migrants awaiting decisions, incarcerating asylum seekers deemed to be at risk of absconding, issuing papers necessary for visa applications, exercising discretion with regard to family and marital on-migration, and administrating regionally specific naturalisation laws. The devolution of responsibility from higher to lower governmental tiers has created discrepancies in asylum reception conditions within countries, obfuscated lines of accountability concerning reception conditions and given central states leeway to both distance themselves from the implementation of tough national level legislation and criticise isolated examples of 'harsh' applications of these laws (Weber, 2003).

Alongside public sector delegation to sub-national state authorities sits the enrolment of a wide range of local non-state actors in immigration control (Cohen, 2002; Coleman, 2009). The primary means by which individuals such as estate agents, doctors and nurses, teachers, driving licence issuers, university lecturers, police officers and private security staff are enrolled in migrant policing is through information verification and collection procedures that they are increasingly required to carry out. So, for example, since 2004 the British National Health Service (NHS) has not been at liberty to treat asylum seekers who have reached the end of their legal process for more than emergency procedures in the United Kingdom, meaning that their status must be checked, reported and recorded by NHS authorities, constituting an important information resource for deportation and removal enforcement teams (Hargreaves *et al.*, 2005). Sigona has identified the same mechanism operating through the responsibility placed on 'social workers to assess failed asylum seekers' entitlements to access support ... and by asking schools to cooperate with [the British state] on parents who do not comply with immigration controls' (Sigona, 2010). Cohen observes that these requirements render 'the statutory dispensers of community care ... investigators of immigration status and withholders of such care from those without the appropriate status' (Cohen, 2002, p. 534).

What is notable about the delegation of responsibilities to local actors is the concomitant closure of individual moral leeway. Where once doctors, teachers and others might have been at liberty to help or turn a blind eye, the creeping ubiquity of the requirement to perform immigration checks has impinged upon the moral room for manoeuvre of the individual. To turn a blind eye now, or to help someone who does not have relevant status on the basis of empathy with the individual concerned or one's moral convictions, is to risk personal sanction for not carrying out the requirements of one's professional position. The alignment of amorality with rational self-interest is therefore again in evidence. Systems that make it rational or necessary for individuals to overlook the suffering of others in the pursuit of their own survival or self-interest risk conjuring the same configurations of personal mores and suffering that led to such widespread misery during the last century. The delegation of mandatory immigration status verification and checking responsibilities has precisely this effect.

A similar risk characterises the 'outward' exteriorisation of border controls. Private transport companies have been incentivised to carry out their own immigration checks and enforcement procedures through the introduction of financial penalties upon companies found to be transporting clandestine immigrants. In these ways, states 'circumvent constraints imposed by juridical and civil rights groups, which may be present at the national or international level' (Lahav and Guiraudon, 2000, p. 64), because transport companies can act pre-emptively – in transit or at source – and thus carry out their activities beyond the scrutiny of these groups. Employers found to be employing illegal immigrants also face financial penalties, as well as prison sentences for employing people without adequate immigration status in various countries. In Britain, landlords have recently been added to the list of groups who face penalties if they deal with migrants without status.

Overlaying the upward, downward and outward eschewing of responsibility for migration is an emergent culture that nurtures adiaphorism. The degree of mediation between the issuing of a border control order and its implementation has become so pronounced that Europe's migration regime has been characterised as a *virtual* realm wherein imperatives, logics, norms and directives are often *never even traceable* to a single source (Bigo and Tsoukala, 2008). Market logics have been introduced through the introduction of tendering for contracts by private corporations and countries who must compete to win them by demonstrating their aptitude for precision and clinical efficiency, which leaves no room for sentimental sensitivity. And the development of border control as a managerial problem has required the introduction of specific expertise and forms of training, a risk-centred approach to migration that merges concerns over terrorism with the containment of people on the move, and the depoliticisation of concern over border control to technical issues of best practice, partnership, norms, standards and regulations (Andrijasevic and Walters, 2010).

Conclusion

In this chapter I have argued that the bureaucratic system of managing borders increasingly precludes literal proximity, and hence encounters, between bureaucrats and their subjects. Such encounters, as a range of moral theorists, geographers and psychologists have noted, hold the potential to move bureaucrats to act not according to adiaphoric principles of government, but according to particular moral sentiments prompted by the needs and demands of specific individuals. Administrative systems of rule are averse to these encounters owing to their disruptive potential, and therefore any reorganisation that reduces the frequency of their occurrence and opens moral distance is unlikely to be resisted by bureaucratic managers.

In the case of border control, moral distance has been opened (following Bauman) by increasing the literal distance between decision makers and bureaucrats, increasing the layers of middlemen or technologies that separate the two, and implementing systems of (dis)incentivisation that promote the rational pursuit of self-interest over moral concerns. This tendency towards keeping people apart is common both to the hierarchical bureaucratic forms of administration that Weber and Bauman described and the governmental states that Jessop has more recently characterised. Indeed, moral distance between subjects and decision makers is facilitated, rather than undermined, by governance arrangements, which increasingly mediate between cause and effect.

The result is that the sort of encounter between Nataliya and system managers like Mr Brimelow that occurred in the Foreign Office in London in 1945, or anyone with personal discretion over individual cases, is arguably even less likely to occur now than it was then. System designers as such may not even exist, as modern immigration control systems tend to emerge from the interactions of various groups rather than through centralised planning. But where elites do exist, they can rest assured that they are shielded by more layers of technology (supposedly more intelligent) than that available in Mr Brimelow's time, and that accompanying this technology are an army of functionaries whose jobs involve dispassionately determining Nataliya's claims by rendering them legible in terms of procedure and routinised decision making, the parameters of which they have neither the authority nor incentive to alter.

We need an approach to 'the state' that is sensitive to interpersonal relations and their suspension. To abstract from interpersonal matters as Jessop (2002) does is to overlook an important consequence of the evolution states have recently undergone. It is not my intention to contribute towards 'state-phobia' (Foucault, 2008): I do not want to give the impression that the tendency towards dehumanisation and depersonalisation of migrants, and the avoidance of encounters, is always pre-mediated and meticulously planned. But just as we might not be impressed by the way water exactly fills the available space in a cup, so we should not be surprised by the 'amazing sophistication and complexity of bordering practices' (Bialasiewicz, 2012, p. 843) that emerge as bureaucracies take the paths of least resistance available to them.

Nevertheless, I also do not want to give the impression that the discussion in this chapter gives a complete account of how decision makers are nurtured to act dispassionately, insensitively and indifferently within the British immigration control system. Although the distance between decision makers and subjects that governance arrangements introduce to border controls forms an important part of the story, various conspicuous questions remain unanswered, including how dispassion is maintained in relation to the thousands of asylum seekers who do arrive Britain and are not kept at arm's length. Are migrants the only ones who are 'distanced' for instance? And why do decision makers who do come face to face with migrants not all react with the same levels of compassion and concern as Mr Brimelow did when he encountered Nataliya? Meetings like these occur when immigration officials assess migrants' cases for asylum via interviews and legal hearings, after they have reached the UK, but only rarely can they be characterised as morally demanding encounters. In the next chapter I will critically enrich the broad, Baumanian account of the relation between moral distance and contemporary border controls presented in this chapter by drawing on empirical material to explore these questions.

Notes

1 This account is taken from Nicholas Bethell's (1974) book, *The Last Secret: Forcible Repatriation to Russia 1944-7*, and uses his pseudonyms.
2 Here I take partialism to mean the view that it is reasonable to show preferential treatment to a particular group.

3   Matthew Gibney's (2004) book-length treatment of the ethical viewpoints of cosmopolitanism and particularism, as well as the dilemmas that both these viewpoints produce, is an accessible introduction to this issue in the context of asylum migration.

4   Krznaric (2014) recognises this same potential for surprise in encounters. 'If you bring two people together with different viewpoints and experiences', he writes (Krznaric, 2014, p. 127), 'the encounter between them can create something unexpected and new.'

5   We might also add certain forms of bureaucratic mapping to the list of technologies that achieve estrangement; see, for example, the maps used by FRONTEX to conceptualise migration routes (in Bialasiewicz, 2012, p. 849).

6   When Gregory discusses distance, however, he departs from a simple account that equates distance and indifference. While he does discuss the idea that the drone represents an extreme example of the separation of cause and effect and that this separation fosters a 'Playstation mentality' that makes it easier to kill (citing Grossman, 2009), he complicates this account by pointing to the immersive, and hence intimate, nature of drone operation (citing Chamayou, 2013). Drone operatives sit only 18 inches from the screen that depicts the consequences of their actions on the ground, making it difficult to make the case that they are detached or removed by the technology they are using. The result is that distance can no longer be relied upon to make violence more abstract and impersonal: rather, Gregory argues, the new technologies that facilitate drone warfare are indifferent to 'near' and 'far'.

7   See Gill (2010) for a fuller statement of my position regarding the concept of the state in research into forced migration.

8   While I discuss the upward, downward and outwards exteriorisation of border control in this section, in fact recent governance innovations go beyond merely 'exteriorising' responsibilities for borders by often simultaneously extinguishing the possibility that migrants can access asylum or legal support at all. Rather than talking about a 'shift' in responsibility to different scales then, a renunciation of responsibility may be a more accurate metaphor (see Andrijasevic, 2010, for a fuller discussion).

Chapter Three
Distant Bureaucrats

In early June 2005 the then-Minister for immigration, Tony McNulty, announced the abandonment of a plan to develop a network of large asylum accommodation centres in rural areas throughout England (BBC, 2005a). The scheme was originally conceived in 1998 and the legislation was passed a few years later under the Nationality, Immigration and Asylum Act (Great Britain 2002). The accommodation centres were to be mandatory, and although asylum seekers would be able to come and go during the day they were to be located in remote areas such as old airfields and 'in one case, where the diseased carcasses of tens of thousands of sick cattle had been incinerated' (Webber, 2012, p. 151). They were intended to speed up the system of asylum determination by reducing decision times, preclude perceived opportunities to work illegally, or engage in housing or financial fraud (Polese, 2013), and remove asylum seekers from local communities from which it was difficult to deport them. The intention was to house around 3000 asylum seekers in four centres of up to 750 asylum seekers each.

In fact, the size of the proposed centres proved to be the sticking point for irate local communities, members of which were concerned about the criminal risks the centres posed to their quiet rural neighbourhoods, as well as the extra infrastructural burden that centres on this scale represented. So although asylum seekers' and migrants' rights campaign groups raised concerns surrounding the suitability of accommodation, provision of adequate health care, capacity of local interpretation services and access to

Nothing Personal?: Geographies of Governing and Activism in the British Asylum System,
First Edition. Nick Gill.
© 2016 John Wiley & Sons, Ltd. Published 2016 by John Wiley & Sons, Ltd.

education for children on site (who were not to be allowed to go to mainstream schools), it was the objections of affluent, middle-class, rural communities that eventually overturned the plans (Hubbard, 2005). The chosen rural areas each happened to be in comfortable Conservative constituencies, which brought the charge that the in-power Labour party was attempting to contain the impact of the centres in opposition areas. Fuelled by a vociferously anti-asylum national tabloid printed press, the plans whipped up such a robust and formidable opposition in the form of local protests, demonstrations and organised legal objections that the planning process became significantly delayed (National Audit Office, 2007). 'In the end', writes Polese (2013, p. 89), 'due to recurring failures to find an appropriate site, the plan for building accommodation centres was eventually dropped by the government … By the end of March 2007, the Home Office stated that about £33.7 million had been spent on the project as a whole [and that] £29.1 million had been recorded as a financial loss'. The accommodation centres would have achieved a level of concentration, isolation and concealment of 'undesirables' in the United Kingdom on a scale not witnessed in Europe since World War II. Yet the cause of the failure of the plans was less to do with moral objections to the centres than with the xenophobia of white middle-class England.

Denied the opportunity to cordon off asylum seekers in specialised 'para-sites' (Serres, 2007) such as these accommodation centres, the Labour government of the late 1990s and 2000s was not short of alternative means by which the remoteness and isolation of asylum seekers could be engineered. A system of dispersal was introduced in 1999 that was designed to 'spread the burden' (Robinson et al., 2003, p. 164) that asylum seekers represented to local authorities away from London and the South-East of England. A previously decentralised arrangement that allowed asylum seekers to live where they wanted to, which for practical reasons was most frequently London, was replaced by a centrally orchestrated system that shifted large numbers of claimants 'to areas of surplus [housing] in the older industrial cities in the Midlands, the north and Scotland' (Griffiths et al., 2004, p. 27). All seven of the major dispersal areas nationally were in the top 20 most deprived areas in Britain on the Index of Multiple Deprivation (Phillimore and Goodson, 2006, p. 1717). Asylum seekers were relocated on a no-choice basis away from 'family, friends, ethnic communities, services and refugee support groups' (Vickers, 2012, p. 52) and were forced to relocate to areas 'with inadequate social provision [and] a lack of qualified lawyers' (Vickers, 2012, p. 52). The dispersal system was designed explicitly as deterrence against asylum seekers' perceived abuse of the British benefits system (Home Office, 1998; Hynes, 2009).

Dispersal was accompanied by a set of 'increasingly hostile' punitive measures (Griffiths et al., 2004, p. 27). The new system included the requirement that asylum seekers move on from dispersal accommodation 'within 28 days of receiving a final decision on refugee status' (Phillips, 2006, p. 542), meaning that many faced compulsory destitution. Even when they were allowed to claim financial support, asylum seekers were forced to subsist on a level of welfare benefits equivalent to only 70% of the mainstream level of income support, a feature that was avowedly designed to reduce the attraction of the United Kingdom to economic migrants posing as asylum seekers – the notorious but mythical 'bogus asylum seeker'. What is more, support was to be in the form of vouchers rather than in cash (supermarkets were told they could keep the change as an incentive to participate in the system – see Webber, 2012), which introduced its own stigmatisation of the separated group.

In these ways, the United Kingdom 'focused its efforts on developing policy that exclude[d] asylum seekers from mainstream society' (Phillimore and Goodson, 2006, p. 1715) by employing spatial strategies of relocation and segregation in order to 'physically separat[e] refugees with status from those without, refugees without status from the rest of society, and refugees without status from one another' (Vickers, 2012, p. 52). The consequences amounted to nothing less than 'institutionalised inhumanity' (Webber, 2012, p. 9). Refugees had little control over where they were dispersed and regularly faced vicious racism in poor, run-down, white working-class areas (Boswell, 2003). The housing itself was supplied through contracts by a mixture of private and social landlords who often sought out the cheapest, semi-derelict housing available, resulting in many asylum seekers living in 'decrepit, unhygienic conditions' (Webber, 2012, p. 93).

Additionally, the award of large contracts to supply housing in poor areas artificially distorted the housing market in these areas. Asylum seekers – who were not allowed to work, were highly visible due to their different appearance, and placed a perceived strain on health and education services locally – were now also seen as responsible for pushing up house prices and rents in deprived neighbourhoods (Phillips, 2006). Families, who often had to walk for an hour to the nearest supermarket participating in the voucher scheme, felt exposed and threatened and began to experience high rates of depression and mental health problems (Patel and Kelley, 2006, pp. 5–6). Women faced an increased risk of domestic violence and, given the levels of destitution, were sometimes unable to look after their health appropriately following childbirth (Chantler, 2010; Webber, 2012). Eventually the calibrated discomfort (Darling, 2011a), alongside racial tensions that culminated in vigilante style violence including the highly publicised murder of a Kurdish asylum seeker in Glasgow in 2001, forced many asylum seekers to flee their dispersal accommodation and return to London and the South-East, losing their right to housing and consequently facing destitution. By 2006 the Red Cross estimated that they had assisted 36,000 destitute asylum seekers, mostly through soup kitchens and night shelters in London (Taylor and Muir, 2006). By 2008 the Mayor of London, Boris Johnson, called for an amnesty for immigrants without status in London, estimating that such a move could generate an additional £3 billion in revenue from taxes and increased wages (British Red Cross, 2010). By 2013, when he reissued his appeal, the situation remained unchanged (Mason, 2013).

For all its inhumanity though, dispersal has not succeeded in removing asylum seekers from the 'horizon of…daily life' (Bauman, 1989, p. 189) in the way the accommodation centres were intended to. Although the failure to launch accommodation centres in the United Kingdom was at least partially compensated by the excision of asylum seekers from their own communities of support through dispersal, urban areas such as Birmingham, Bristol, Cardiff, Glasgow, Leeds, Liverpool, Manchester and Sheffield have adapted to the dispersal arrangements and now, although dispersal is by no means an easy option, there are strong communities of solidarity in these and other major British cities. The work of communities of solidarity such as the Cities of Sanctuary movement in the United Kingdom, the National Coalition of Anti-Deportation Campaigns (now called Right to Remain) and No Borders has rendered at least some asylum seekers ineffaceable, as evidenced by the steady stream of successful attempts to resist deportation.

Faced with this inextricability, my argument in this chapter is that the immigration control system in the United Kingdom has bolstered its spatial management of asylum seekers within the country with a programme of spatial management of its own decision makers. This argument rules out the easy assumption that moral distance is primarily concerned with distancing subjects from decision makers and not vice versa. Given the bureaucratic aversion to proximity between decision makers and subjects, which can lead to troublesome encounters between them, and given the failure of plans to obscure subjects in remote rural accommodation centres, decision makers themselves have been the targets of a sustained process of estrangement from the very asylum seekers that they make decisions about on a daily basis. This story is less frequently told than accounts of dispersal and its consequences yet results in the same outcome of keeping people apart.

The chapter details the spatial stages through which asylum sector decision makers became estranged from their subjects during the late 1990s and 2000s. I argue it is possible to distinguish three discrete stages in the distancing process beginning with the *insulation* of decision makers from contact with their subjects, followed by *buffering* them by positioning other organisations between them, and culminating in the splitting of the decision-making group itself in order to establish *competition* among separate units over the most efficient way to achieve abstract objectives. Two points follow. First, it becomes clear that decision makers are as much subject to the spatial strategies of asylum seeker management as asylum seekers themselves. Their rule is as much a challenge as that of their subjects. Second, the consequences of these stages are as much psychological in rationale as they are spatial: I identify the systematic spatial management of decision makers as a key element in the production of psychological dispositions amenable to dispassionate bureaucratic rule.

The National Asylum Support Service

The introduction of dispersal in the United Kingdom was accompanied by the creation of a specialist agency, the National Asylum Support Service (NASS), which was a self-contained arm of the Home Office. Prior to the establishment of NASS, asylum seekers were able to access the mainstream welfare benefits system in the United Kingdom, but in 1996 the Asylum and Immigration Act (Great Britain 1996) curtailed that access. The result was that by 1999 local authorities were having to meet the needs of otherwise destitute asylum seekers, often through ad hoc arrangements. Although this provision was chaotic and under-resourced, lawyers and activists were later to look back upon this period as one in which at least some of the local authorities in the United Kingdom took direct responsibility for the support needs of asylum seekers. One immigration lawyer who had practised in Bristol throughout the 1990s recalled that 'we just dealt with the local council social workers. There was a special department that just dealt with asylum seekers so I got to know them very well and everybody had their own designated social worker'.[1] Local authorities, however, felt that they had been drawn into the provision of support without adequate consultation and therefore provided it only grudgingly. Although they were legally compelled to provide support to tens of thousands of asylum seekers, they argued that they were receiving 'disproportionate and unsustainable demands' (National Association of Citizens Advice Bureaux, 2002a, p. 7) and that they had neither the financial capacity nor the competence to adequately support the asylum-seeking population, especially in the South-East of England. Compounding the local authorities' sense of grievance at having to meet the needs of destitute asylum seekers, asylum-seeking numbers were increasing markedly at that time (see Figure 3.1).

NASS was therefore created in order to implement the system of dispersal away from the South-East and, in response to political and media pressure, 'to discourage abuse of a chronically overstretched system' (Noble et al., 2004, p. 50). Figure 3.2 illustrates the effect of the efforts towards dispersal: asylum seekers who were housed by NASS are represented by the white bars, which occur almost exclusively outside Greater London. It was mostly only those asylum seekers who forfeited their right to accommodation and relied on 'subsistence only' support who were able to concentrate in London (represented by the black bars). The key functions of NASS included: (i) ensuring that only those eligible for support received it; (ii) contracting

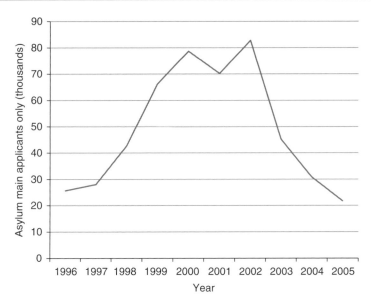

Figure 3.1   Number of asylum applicants to the United Kingdom, 1996–2005. Adapted from The Migration Observatory at the University of Oxford (2014).

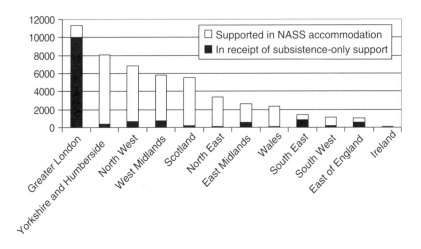

Figure 3.2   National Asylum Support Service (NASS) supported asylum seekers by region, 2004. Adapted from North East Consortium for Asylum Support Services and North of England Refugee Service (2004).

with private accommodation providers to provide a sufficient supply of suitable accommodation for those who needed it; (iii) facilitating the financial support of asylum seekers, working in partnership with a range of community stakeholders; and (iv) providing emergency accommodation when required.

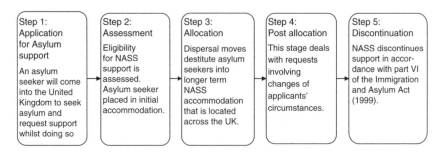

Figure 3.3   End-to-end National Asylum Support Service (NASS) process. Adapted from National Asylum Support Service (2007).

By 2002 NASS was supporting 82,000 asylum seekers (Noble *et al.*, 2004) and the total annual cost of the work NASS commissioned exceeded £1 billion (Noble *et al.*, 2004). Figure 3.3 shows the end-to-end process that NASS oversaw. The flow chart begins with the receipt of a written application for asylum support, and it was NASS employees who decided whether particular individuals were eligible for support.

Yet NASS as an organisation, and consequently its employees, were charged with incommensurable objectives. As an independent report reviewing the operation of NASS noted in 2003:

> Staff in NASS…have to wrestle constantly with an inherent contradiction in their role. On the one hand they are providing a welfare support system to some very vulnerable people. And at the same time they are working within a framework of deterrent-based policies and legislation… [G]etting the balance right in any individual case is not easy, and the challenge it presents to junior staff in particular should not be underestimated
>
> Noble *et al.*, 2004, p. 8

The confusion between the imperatives to support and investigate asylum seekers was palpable amongst NASS employees. In the summer of 2005 I conducted research at the South-West NASS office in Portishead, near Bristol, and was able to interview staff and access the back-offices of the building[2] (see Appendix for methodological details). The bulk of the work employees carried out involved determining the eligibility of cases for support, making sure eligible asylum seekers were supported by responding to changes in their circumstances and negotiating a large number of contracts with a variety of agencies to provide all aspects of the support NASS made available. Contracts had to be negotiated with housing management companies, voluntary sector organisations (to provide emergency accommodation and help asylum seekers complete the application for NASS support), and coach companies (to transport asylum seekers to their allocated accommodation). Commenting upon contradictions in their role that ran throughout these activities, one employee described the temptation to de-emphasise the supportive side of his work. 'Where NASS are concerned,' he explained,

> we are responsible for sending out the money and the accommodation, and I think there is an element of NASS that sees that as the end of the support bit. We also oversee the monitoring, the checking, the gate keeping bit and sometimes I think that that actually becomes more important than the actual supporting[3]

Interviewees told me that many of their colleagues had applied for jobs with NASS because they felt drawn to supporting asylum seekers in some capacity. Indeed, some had worked for voluntary refugee support agencies in the past. Yet one employee involved in housing management expressed his misgivings about the sort of work he was expected to perform. 'I try to be professional and fair wherever I can be, but it's like I am playing God with people's lives', he complained. 'I'm personally against the BNP [British National Party], *Daily Mail*[4] attitude, but I'm often put in an awkward position.'[5] He gave the example of being asked by the police to relocate a man who was perceived to be the cause of ethnic tension and antisocial behaviour in a small town in his dispersal area. 'There was no evidence', he recalled, but anticipation of 'trouble' among the police was enough to transfer him from his community in the South-West to a different location in the North of the country. The transfer was against the man's wishes because he and his family had built up social networks in the town. My interviewee was left in no doubt about the punitive and potentially racist nature of the transfer. 'They couldn't charge the person with an offence', he explained, 'so they made us move them to Manchester instead.'

Another employee outlined her uneasiness at the way the supportive functions of NASS had to make way for investigative imperatives. Her role included investigations into allegations of domestic violence and taking action, wherever necessary, to protect asylum-seeking women. But she had also recently been asked to look for any signs of illegal employment in the households that she visited (asylum seekers were usually not allowed to take paid employment). This, in her view, produced 'a serious trust issue and women will not come forward'.[6] To make matters worse, her visits were also being channelled away from women and families. Instead, so-called 'outreach' visits were to be targeted at young single men, because these were 'the ones most likely to be working illegally'. The frequency of visits to other types of household, including women with children but no male partner, were consequently scaled back because they 'would not be able to work illegally due to their responsibilities', despite the fact that they might be the ones in most need of support. As a result of the systematic reduction in the ratio of supportive to investigative activities, she expressed her consternation at having to be 'constantly hardened' in the role.

Insulation

Faced with confusion and ambivalence amongst its staff surrounding the contradictions between their roles as carers and their investigative responsibilities, NASS undertook a series of steps that disentangled decision makers from their potential attachment to asylum claimants. Crucially, the decision was made not to introduce NASS-run service counters for asylum seekers at the inception of NASS in 2000. As the review of NASS operations in 2004 noted, 'NASS was set up as a self-contained operation...so that it could concentrate on its intensive processing and contracting activities... At the time, IND [Immigration and Nationality Directorate] were trying to deal with significant case-working difficulties and there was merit in setting up NASS in a way that insulated it from the downdraught of those problems' (Noble et al., 2004, p. 50). This meant that 'NASS quickly established...an unenviable reputation for administrative inefficiency and bureaucratic inaccessibility' (National Association of Citizens Advice Bureaux, 2002a, p. 1).

Applications for NASS support, for example, had to be made through a standardised form (see Figure 3.4) rather than in person. The 24-page form constituted part of what Winder (2013, p. xi) has called the 'elaborate paper barricade' that the United Kingdom and other developed countries have constructed in recent years to regulate and control migratory movements.[7] It represents a thorough interrogation that dwells in detail upon highly personal aspects of the applicant's case and situation, including the value of their land and whether it can be liquidated in order to support them, the value of their assets such as jewellery, and the value of electrical goods that they own such as TVs and DVDs, obviously carrying the implication that these are to be sold in order to support the applicant. Despite its length though, the form obscures the complexity of asylum cases. It mobilises a detached, sterile account of the asylum seekers it describes, distilling 'facts' from emotions, abbreviating long histories of often arduous travel, and curtailing accounts of loss and suffering to small blue boxes on a page.[8]

Partly as a result of the difficulty of navigating the application process, a report published by the National Association of Citizens Advice Bureaux (2002a) issued a desperate plea for NASS to improve its accountability. The report provided details of countless glitches and errors that would routinely occur in the process of matching asylum-seeking families to accommodation, delivering them securely from one location to another and making sure that their financial support reached them. Usually the problems were simply procedural in nature such as vouchers being sent to the wrong address, financial provision to attend immigration hearings not arriving, letters not being sent or not being picked up, or confusing guidance being issued by NASS. The problems were compounded by bureaucratic incompetence. At one point, for example, NASS sent out thousands of letters to its claimants containing the instruction 'If you have not received this letter you should

Home Office
BUILDING A SAFE, JUST
AND TOLERANT SOCIETY

National Asylum Support Service Application Form

Land in the UK?   Yes/No

If yes what is the value of land in the UK?

Currency?

Can you liquidate this asset?   Yes/No

Land outside the UK?   Yes/No

If yes what is the value of the land outside the UK?

Currency?

Can you liquidate this asset?   Yes/No

Valuable jewellery   Yes/No

If yes what is the value of the jewellery?

Currency

TV, DVD, Electrical Goods   Yes/No

If yes what is the value of the good?

Currency

Car, Vehicle   Yes/No

If yes what is the value of the good?

Currency

Figure 3.4   Graphic illustrating the structure and tone of the Application Form for NASS support, based on pages 1 and 7. Adapted from the National Asylum Support Service Application Form used by the Home Office in 2012.

contact NASS immediately on the number above' (National Association of Citizens Advice Bureaux, 2002a, p. 3). 'One can only imagine the confusion of such individuals on receiving a letter from NASS – the governmental body charged with meeting their welfare needs – that concludes with th[is] statement' (National Association of Citizens Advice Bureaux, 2002a, p. 3).

However, as one volunteer with a local refugee supporting charity in Bristol noted, there was a deeper issue of an inward-looking culture throughout NASS. Without front desks that asylum seekers could visit in person NASS 'appear[ed] to be abdicating responsibility for that face to face sorting out of just the little errors and things that creep into these things', she explained. 'I mean we were busy enough before and it seemed as if NASS were just handing on problems to other agencies.'[9] Her concerns resonated with the conclusions of the Citizens Advice Bureaux report:

> the principal issue here is that NASS has no local counter (or 'drop-in') services... This lack of local access points has been seriously compounded by a...stakeholder averse culture within NASS [and] a paucity of up to date, accurate guidance on how to contact NASS or otherwise resolve problems. ... [G]iven effective access to responsive, local NASS counter or 'drop-in' services, most...problems could be resolved by asylum seekers themselves, without outside intervention.
>
> National Association of Citizens Advice Bureaux, 2002a, p. 4

The consequences of the accessibility deficit of NASS were sometimes inexcusable, with mothers left without sufficient money to sustain breast-feeding, and others having to endure days of hunger before their allowances arrived. As just one example, NASS introduced a system of emergency tokens in February 2003 for any claimant whose support had been interrupted. The tokens were to be delivered by courier 'within 48 hours' (National Association of Citizens Advice Bureaux, 2002b, p. 24) of notification of the interruption reaching NASS, and needed to be signed for by the recipient. NASS stipulated that recipients must remain at their address for the whole 48 hours in order to receive the support. Unsurprisingly, after just a few weeks, NASS found that nearly 50% of the tokens were undeliverable at the first attempt, because 48 hours, often over three days, is too long to expect hungry people to stay indoors waiting for a delivery that they suspect may never arrive. This sort of ineptitude brought the charge that the system of support was being run for the convenience of NASS and that NASS was incapable of empathy with its largely non-English-speaking client group (National Association of Citizens Advice Bureaux, 2002a, 2002b).

In response to some of these criticisms, NASS made half-hearted attempts to improve accountability. Phonelines were introduced so that claimants could make enquiries directly to a NASS employee, and the regional offices were strengthened to improve accountability. In both cases, though, the innovations actually did little to improve accountability and in some respects undermined it further.

The Phonelines

In the case of the phonelines, NASS set up dedicated telephone helplines to take queries about its services. However, the helplines were roundly criticised. Callers described the unhelpful telephone manner, the experience of giving details about problems only to be told that they would be called back by someone else at an unspecified time (and then never receiving the call), long waiting times before a call was answered, and no record of previous communication held by NASS. At times the difficulties of contacting NASS descended into farce. Citizens Advice described the difficulty of using faxes to contact NASS, for example:

> Given the continuing difficulty in contacting NASS by telephone, CAB advisers commonly send urgent letters to NASS by fax, using fax numbers set out in the NASS telephone directory and other guidance posted on the Home Office IND website. … [H]owever, NASS managers [have] advised Citizens Advice that fax machines at NASS are not routinely monitored for incoming faxes, and asked that faxes should *not* be sent to NASS unless the sender has first telephoned NASS to confirm.
>
> Citizens Advice (2003)[10]

Volunteer refugee support workers in Bristol described the 'incredible problems trying to get through to NASS. If somebody came in and they had not had their vouchers or whatever we'd ring and be on the line for ages and ages and just not get through or you'd be passed to a lot of people, so you'd wait for ages and then be passed to another department and they'd say "oh you should have spoken to another department".'[11] The independent review of NASS concluded that 'the system for handling telephone calls is…just not fit for purpose, and sorting that out has to be a key priority' (Noble *et al.*, 2004, p. 8),

> The telephone call-centres are the main communications route into NASS. They are used by asylum seekers and the voluntary sector agencies that help them, to try to sort out problems that are often very urgent [but] the present level of performance in handling and responding to telephone calls is just not acceptable.
>
> Noble *et al.*, 2004, p. 9

Incredibly, the experience of the regional employees themselves in getting through to their colleagues in the central office in Croydon, London, was much the same as that of external volunteers. They often had to rely upon the same helpline, which resulted in all manner of problems. 'It seems crazy', one employee told me, 'to have a central body when you've got people dispersed all over the country and they're having to try to get through on the phone lines which are engaged most of the time...it's madness.'[12] Another employee complained that,

> They [meaning his NASS employee colleagues in Croydon] will not give us the names of people dealing with cases, even to the regions. They are 20 years behind. They will put the phone down on you. I called one guy on an extension, he gave me another extension number but when he picked it up it was clearly the same guy, and he just said the person I wanted was out for lunch.[13]

Another interviewee lamented the high turnover of staff in Croydon where the central NASS team were located, their general inexperience and their lack of personal accountability. 'They think that we [in the regional offices] are taking work from them, so they don't want to help us at all', he complained. 'They are arrogant on the phone and often the phone numbers are wrong.'[14] Various authors have noted the internal contradictions inherent to states. Painter (2006), for example, notes the prosaic nature of state formation and urges that 'our accounts...give full weight to the heterogeneity, complexity and contradictoriness of [the] state' (Painter, 2006, p. 764), and Jeffrey (2013) has highlighted the need to develop a 'critical stance that challenges the ontology of the state as a coherent set of institutions' (Jeffrey, 2013, p. 23). In studies of the bureaucratic management of border control, Heyman (1995) has highlighted the reliance of the United States Immigration and Naturalization Service upon the 'bureaucratic thought-work' (Heyman, 1995, p. 261) and 'world-views' (Heyman, 1995, p. 261) of individual officers to make sense of, reconcile and piece together its contradictory policies. And in her ethnographic study of the Canadian response to human smuggling, behind the projected facade of coherence and control of the state, Mountz (2010) describes the panic, crises and 'fascinating set of power struggles within the state' (Mountz, 2010, p. 61).

The experiences of the regional NASS employees bear out much of this confusion, contradiction and chaos. 'Communication with Croydon is infuriating', one management-level employee exclaimed. 'They cut you off. People refuse to give their name and often they are temps.'[15] The employee with responsibility for supporting victims of domestic violence in NASS housing had a similar experience. 'I have lots of trouble reaching Croydon', she recounted, 'and even when I don't get through the phoneline doesn't let you leave a message, then it just goes dead.'[16] While most of the difficulties were traceable to incompetence and a lack of resources, at times miscommunication within NASS was more calculated than accidental. One manager who helped to organise deportations and removals described her tactic of not informing her colleagues within NASS of the plans of her department:

> NASS should be informed when removals are imminent, but sometimes they aren't. So there are meetings three weeks ahead to detail the plans, but NASS aren't always invited. Sometimes the [private accommodation] providers that have been contracted by NASS are present, but not NASS themselves. So there are some very awkward moments and you really have to keep on top of who knows what.[17]

The reason she gave for this sort of secrecy was that NASS employees might reveal the timing of the arrest and deportation attempts when dealing with 'clients'. But this strategy did nothing to alleviate the sense in which NASS employees felt isolated and dispirited especially in the regional offices. 'I think we are a buffer between central Croydon and our clients', one worker complained,[18] while another reflected that 'NASS is just too big, and the workforce is very demoralised.'[19] NASS employees reported that 'there's a general perception in the office that NASS regionalisation is not working',[20] while charitable refugee support group employees viewed the NASS regional workers as rather impotent. 'I mean they are very good, they are good colleagues of ours', one volunteer explained,[21]

> but they don't have an awful lot of power. They can come out and look at local housing conditions and that sort of thing but if we have somebody who applied for [financial support] months ago and still hasn't had a reply there's not much they can do about it. We find that we often find out what Croydon are doing before [they] do, so their hands are very much tied.

The phonelines, then, achieved a two-fold isolation of regional NASS employees. First, it meant that NASS employees were denied the opportunity to meet the asylum claimants that they made decisions about, and had to rely instead upon an impersonal and standardised form to inform their decisions. And second, because the phoneline was an important way for regional NASS employees to communicate with central employees in London, they simultaneously isolated regional from central workers.

Regionalisation

Another response to the charge of poor accountability involved strengthening the regional offices. Facing an avalanche of complaints about the inadequacy of support, the decision was taken to regionalise NASS further by expanding the regional offices so that they could take on some of the work of contracting local housing providers and liaising with stakeholders

(Noble *et al.*, 2004). By 2005, the functions of the 12 regional offices included the management of housing contracts, making sure asylum seekers had arrived safely and that their accommodation was of the required standard, investigating cases of illegal working by NASS-supported applicants, investigating antisocial behaviour either experienced or perpetrated by asylum seekers, and liaising with regional stakeholders including local authorities, local police forces and the local voluntary sector. According to the NASS website, the idea behind the comprehensive regionalisation of NASS was to increase the Service's proximity to local actors and NASS users. 'By enhancing its representation in the Regions, NASS will be closer to its partners, stakeholders, and its customers and consequently will be able to provide a stronger, more targeted and effective service', the Immigration and Nationality Directorate (2005) boasted.

Yet my research at the South-West NASS office indicated that, if anything, the regionalisation drive achieved the opposite effect. The South-West regional office was located in Portishead, a town with a population of around 18,000 (see Figure 3.5). Portishead is a seaside resort and a retirement town. One resident described it in the following terms:

> The housing is not cheap and I would say it is predominantly middle class [and] white as well. So that's the principal make-up of the population here. Although there has been increasing diversity of population in Portishead, that diversity I would say is still white European.[22]

Another characterised it as having 'quite a large Conservative faction ... who haven't really been comfortable with any changes in the town.'[23] The town has expensive housing and as a result suffers from a missing generation of 20- and 30-something-year-olds who tend to have to locate in nearby Bristol where the housing is more affordable. In terms of the demographic make-up of Portishead in relation to Bristol there are consequently some stark dissimilarities. 'Although we're only ten miles from the centre of Bristol,' one respondent explained,[24] 'in a cross cultural sense we're on a different planet, because Portishead is mainly mono cultural.'

Like many British seaside resorts, Portishead is a place that is deeply resistant to outside influences (see Burdsey, 2013, for the general case). Portishead does not get much national news coverage, but did make the headlines in April 2004 due to its staunch opposition to the location of an asylum screening centre near to, and in addition to, the existing NASS office. The proposal would have seen asylum seekers visiting the town in order to register their details, provide fingerprints and be issued with registration cards. When a public meeting was called in Portishead to discuss the proposals, however, opposition was so staunch that it attracted national attention. 'The hostility...was tangible. ... Any attempt to speak in favour of the centre was shouted down', reported *The Observer* (Bright, 2004),

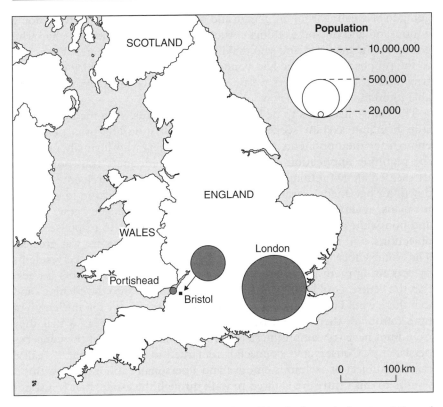

Figure 3.5   Map of Portishead and Bristol. Source: Exeter University Geography Department Map and Print Room. Population figures correct July 2014.

a national newspaper. The objections centred around an exclusive housing estate, The Vale:

> The Vale estate is neat and bright, the location of many a dream home. But as you snake through its traffic-calmed roads and cul-de-sacs, you glimpse a modest office block on its fringe. That is where the Immigration Service wants to bring asylum seekers. The asylum seekers won't be staying in the area. They will be coming into Portishead for a few hours and then going back to their temporary homes elsewhere in the South West. But many local people are furious, fearing what the arrival of 50 strangers a week will mean.
>   BBC News, 'How Portishead Divided Over Asylum', 21 April 2004.
>   Reporter: Dominic Casciani

This fear seemed to know no bounds. One church leader who attended the public meeting about the centre detailed the sort of objections that were raised. 'All the classic stereotypes were mentioned', he recalled,[25] 'my daughter's going to be raped as they walk from the bus to the centre, they're going to be stealing from my garage and my house, they're going to be doing a thousand and one other things'. Another attendee described a 'chap who got up [at the meeting] and said that he lived in the housing area fairly close to the proposed centre, he had young children who like to play out in the street and if asylum seekers were to come to the centre they wouldn't be able to do that because they would be in danger'.[26]

The centre itself was to be located only 250 yards from the nearest bus stop bringing asylum seeker visitors from Bristol. This was not close enough for the opponents of the centre, however, who formally opposed the planning application on the grounds of inadequate transport links. 'It seemed an awful thing to have to say, that they're not welcome to walk down our pavement', recalled one resident who saw no reason to object to the plans, 'it almost brought back images of South Africa, you know? Whites and non-whites and I just felt very saddened by that.'[27] In response to the objections, a new bus stop only a few metres from the centre was erected. The Home Office issued a statement aimed to quell the fears of local residents by reassuring them that there would be no more than 50 visitors a week, that the centre would not be open at the weekends or into the evenings and that there would be nowhere for the asylum seekers to stay for any length of time (although there would be a small indoor waiting area so that they would not have to wait on the street in an unsightly way). Yet the concerns persisted. A Conservative counsellor remarked to the press that: 'I realise that the subject of asylum is emotive and I personally abhor bigotry. But if visitors to this centre are allowed to walk through the estate, that is just not on' (Casciani, 2004). 'People living at The Vale are concerned', the same Councillor continued, 'because they have bought their new homes not expecting that asylum-seekers will be walking around them. It is the biggest purchase of their lives and they are worried and aggrieved' (Thompson, 2004). Another woman expressed her misgivings to a local reporter: 'I'm not against them coming here as such. But they should not be permitted to wander around the country without some kind of regulation' (Casciani, 2004). *The Observer* and the Home Office – the latter not known for its liberal views – were astonished by the robustness of the opposition in the town.

> Home Office officials have been taken by surprise by the strength of feeling in Portishead. Previous campaigns [against accommodation centres elsewhere in England] have centred on government plans to build large residential centres for asylum seekers in rural areas where local concerns involved an influx of large numbers of outsiders. ... But not a single refugee is going to live in Portishead.
>
> Bright, 2004

Eventually the office was built, although not before the government at the time was forced to override the findings of their own consultation with local residents (originally intended to assuage concerns) in order to build the centre. For its part the NASS office continued to be located in the same office block as the proposed centre throughout the furore. What the episode made clear is that given the public feeling towards asylum seekers in Portishead, expanding the NASS office at the expense of both the local council offices in Bristol before the creation of NASS, and the central NASS offices in Croydon in the few years following its creation, can hardly be seen to be moving NASS employees closer to their 'customers'. As one employee at the NASS office remarked, 'decentralisation is a way to separate responsibility from authority. The centre is cutting budgets and making changes without having to face responsibility for the backlash.'[28] Indeed, during the intense public debate about the siting of the screening centre, NASS staff themselves faced 'a degree of anger directed against them', including 'frosty looks and offishness'.[29] These are not the actions of a community that was likely to generate opportunities for decision makers to become 'closer' to asylum claimants, as per the stated aims of the regionalisation drive. On the contrary, Portishead showed itself to be aggressively white and exclusive.

What regionalisation in fact achieved, was the relocation of decision making to sites that were culturally more distant from asylum seekers than either the dispersal areas themselves or the NASS central offices in Croydon. In this way NASS nurtured the systematic insulation of employees from contact with their subjects, in this case by locating them in white, monocultural locations 'on a different planet' from those living in the no-choice NASS housing itself. If accommodation centres had failed to remove asylum seekers from the daily lives of decision makers then the phonelines, the lack of front desk functions, the form-based paper applications and the duplicitous regionalisation process succeeded in removing decision makers from migrants' everyday lives instead.

Buffering

The role of third-sector organisations in resisting or facilitating this process was contested throughout this period. Third-sector groups operating in the asylum support sector are heterogeneous to say the least, and generalisations are difficult. They range from small, issue-specific, informal organisations to large charities, and can be refugee-led or not, motivated by different concerns and displaying very different attitudes towards working alongside government agencies (Gill et al., 2012). But a key element of the inception of NASS was the contracting of charitable agencies to provide 'one-stop' advice services in each of the main regions of the United Kingdom. Although NASS lacked front-desk functions of its own, these agencies were to be the face of the dispersal system. In the first two years of NASS's operations, six organisations (the Refugee Council, Migrant Helpline, Refugee Action and the Refugee Arrivals Project in England, the Welsh Refugee Council in

Wales, and the Scottish Refugee Council in Scotland) received grants totalling £34 million from the Home Office to provide signposting services for asylum seekers, such as help with the application process, help sorting out their claims and help finding emergency accommodation while they waited for NASS to make decisions.

Concerns were raised that charities were allowing themselves to be 'co-opted' (Zetter *et al.*, 2005, p. 173) by accepting these contracts, which came with stipulations such as prohibiting them from supporting refused asylum seekers. 'Whilst publicly denouncing the scheme', Hynes (2009, p. 102) writes, 'there was no sustained campaign against dispersal from the voluntary sector, which ultimately undertook a frontline role in the local level implementation of dispersal.' Briskman and Cemlyn (2005, p. 179) conducted interviews with voluntary agency workers in the early 2000s and concluded that while many '[i]ndividual workers, statutory and voluntary, seek to make a difference' they frequently became 'enmeshed in managing an unsatisfactory situation [in which] provision is under-resourced and uncoordinated, leaving basic needs unmet' (Briskman and Cemlyn, 2005, p. 179).

Concerns in the South-West resonated closely with these criticisms. Refugee Action was the charity contracted to provide signposting services by NASS in Bristol, and while one of their managers could see some logic in the system of contracts that NASS had designed, they were in no doubt about the way the contracts obfuscated NASS's accountability. She explained that:

> When NASS was set up in 2000 they realised that the Home Office don't have a background in advice, giving information and support, so I think they recognised that existing voluntary agencies that had a long track history of working with asylum seekers were in the best place if they were going to make dispersal happen.
>
> NG: Okay, so you don't feel as if they hide behind you to some extent?
>
> They do, yes, without a doubt, without a doubt. I mean they do have regional teams, there is a regional team but it is based out in Portishead![30]

Volunteers in Bristol were concerned that Refugee Action, which had played a leading role in critiquing local policy in the past, was neutered by the contracting arrangements. One anti-deportation campaign group organiser pointed out that Refugee Action 'can't campaign like us: we don't have any strings attached because we don't take money from anybody.'[31] In contrast, Refugee Action faced

> a potential tension because NASS of the Home Office will feel 'well hang on we're funding you, you can't then turn round and criticize us', or Refugee Action will feel 'well we can't be too outspoken about government policy because we're waiting for the cheque to come next week and if that cheque doesn't come we can't pay the rent'.[32]

The contracting arrangements also reduced NASS's engagement with organisations operating in the asylum support sector and asylum seekers themselves. The Bristol Refugee Inter-Agency Forum (BRIAF) would meet bimonthly in central Bristol to discuss the challenges facing refugees and asylum seekers in the city. The Forum was run by the refugee community and was well attended, with representatives from Bristol City Council, the South-West Regional Government Offices, Bristol's police force, Refugee Action and a range of advocacy organisations and representatives of the asylum-seeking and refugee communities in the city.

Yet NASS's representation on the Forum became increasingly sporadic from 2002 onwards. When I interviewed one of the convenors of the Forum, he outlined the difficulties he faced in getting NASS along to the meetings:

> They used to bring statistics to show how many refugees there are, are there any asylum seekers, what are the issues, how can we overcome them. So they were an accountable body. However, over the last year and a half or so I've noticed that NASS has been really not that well represented in BRIAF. They weren't present two meetings ago but NASS hasn't been exactly effective recently in any case. They're less eager to engage with BRIAF, going to BRIAF is not a priority.[33]

NASS's lack of attendance at BRIAF denied them the opportunity to engage with the asylum-seeking communities they were seeking to support in the South-West. By not attending BRIAF, employees remained insulated from the concerns of asylum seekers in the city. While the refugee community clearly missed out on the information that NASS used to provide, NASS itself also missed the opportunity to engage with asylum seekers and be accountable to them.

One NASS manager justified her lack of attendance and that of her team by describing both the lack of formalism of the BRIAF meetings and the fact that Refugee Action would alert them to anything significant that was discussed in any case. 'Very little comes from BRIAF', she complained. 'It needs more cohesion, a regular reporting structure and a tighter remit about what it's there for; it needs more continuity.'[34] In response to the charge of lack of cohesion the convenor of BRIAF defended the relatively informal structure of the Forum, however: 'If BRIAF was to become a formalised agency with a constitution and a fixed membership then [attendees] would lose their freedom to come and criticise what their department is doing', he argued.[35] Nevertheless, this was enough for the NASS manager to decide that it was not worth attending in person or sending one of her team. 'In any case', she remarked, 'we will find out what is going on from our partners [meaning Refugee Action].'[36]

In this way Refugee Action risked buffering NASS employees by fielding the worst consequences of the decisions NASS officials took during this period and by acting as a proxy for their involvement in the local refugee and asylum support community. John Lachs (1981, p. 12) writes that the 'most serious consequence of mediated action is the psychic distance it introduces between human beings and their actions. We quickly lose sight of...the long range effects of actions [and] how it feels to cause what we condone'. The positioning of charities like Refugee Action expressly between NASS employees and the asylum seekers and asylum support community they affected made the charities into mediating institutions and opened the door to NASS employees' indifference towards, and ignorance of, the consequences of their decisions.

Competition

The stage was now set for the introduction of techniques that reduced asylum seekers to mere entries on spreadsheets and component figures in overall totals and trends. A document was published by the Home Office in September 2004 entitled 'Review of Resourcing and Management of Immigration Enforcement' (Home Office, 2004), which examined in detail the most efficient and cost-effective means of removing failed asylum seekers and others refused permission to remain in the United Kingdom. The objectives of the report were to make recommendations in order to maximise the total number of removals per member of staff involved in enforcement, and to minimise the total cost per removal as part of a broader drive towards the expansion of deportation in the United Kingdom (see Gibney, 2008, for a discussion of the deportation initiatives undertaken at this time). Among its recommendations were improved 'performance management, priority setting and tasking' (Home Office, 2004, p. 3), 'greater effectiveness and efficiency' (Home Office, 2004, p. 3) and 'more flexible deployment of resources' (Home Office, 2004, p. 3). Managers across the IND and NASS were encouraged to 'rigorously target their activities towards the most productive operational work' (Home Office, 2004, p. 5) and 'collect management information on a regular and systematic basis which enables them to measure inputs against outputs' (Home Office, 2004, p. 5).

Part of this overhaul involved explicitly targeting 'failed asylum seekers who are costly to support (e.g. families)' (Home Office, 2004, p. 47). Asylum seekers who visit reporting centres[37] were also to be targeted, because 'asylum seekers who routinely report to Reporting Centres during all or part of the asylum process may develop a pattern of compliant behaviour' whereas 'an arrest "in the field" of an individual FAS [Failed Asylum Seeker]...is likely to be about four times as expensive as detaining a failed asylum seeker when they report' (Home Office, 2004, p. 22). In effect, targeting of this sort ensured that the most vulnerable and obliging asylum seekers, who honour the demanding schedule of check-ins and meetings, were the ones most exposed to the threat of removal because they were expected to offer least resistance and therefore be cheapest to deport.

NASS was directly implicated in the new proposals, not least in plans to remove families and children, which were seen as an underexploited way to increase deportations and meet targets. 'We note that operational activity related to family removals is relatively low' (Home Office, 2004, p. 25) the report read. 'We recommend that the Immigration Service should examine in conjunction with NASS how they can increase the numbers of successful family operations despite the inherent difficulties' (Home Office, 2004, p. 25). The report also introduced a competitive logic across the regional jurisdictions of the IND and of NASS. 'The use of Reporting Centres to facilitate removals is more effective in London than in the regions', the report noted for example. 'If all Reporting Centres achieved the current London average for removals per reporting event there would be an overall increase in Reporting Centre removals of 20%' (Home Office, 2004, p. 21).

This logic of comparison and competition among the insulated and buffered group, according to such abstract metrics as 'removals per reporting event', 'cost per removal' and 'detections per package actioned' (Home Office, 2004, p. 49) is a key innovation that relies upon the moral remoteness of functionaries. For each office, NASS introduced a 'scorecard' indicating performance against the top 40 performance targets, as agreed by the senior management team, and 'dashboards' containing detailed data for each 'business area' that senior management could read 'at a glance' in comparison to other regional offices.

Some of the key metrics that offices were encouraged to compete over were widely publicised among NASS employees. For example, the reception area and office noticeboard at the South-West office were dominated by correspondence from NASS headquarters in Croydon that emphasised the imperative to reduce the backlog of outstanding claims for support and celebrated the increasing number of 'fraudulent' claimants detected both by the IND and by NASS during the processing of requests for support. One of the pieces of correspondence on display was a 'Regionalisation Newsletter' informing employees at the regional offices of NASS's progress in implementing the drive towards regionalisation. Figure 3.6 is an adapted extract from this newsletter.

The extract compares the average time taken to decide upon a claim for support received by NASS in the various places that host regional NASS offices. The offices were under extreme time constraints to meet the target times for determining cases for support. The time taken to decide upon the level of support an asylum seeker is entitled to was strictly monitored, with a standard target of 55 minutes and a fast-track target of 39 minutes,[38] bearing in mind that the forms are each 24 pages long. Regionalisation introduced a high degree of competition between regional offices on this metric. The Leeds office, for example, received praise in the newsletter for

Team	Standard casework-actual average time taken. [Target 55 minutes].	Fast track-actual average time taken. [Target 39 minutes].
Leeds	24.00	23.17
Glasgow	30.44	27.55
Dover	24.34	29.41
London	47.34	31.38
Newcastle	Training	
Peterborough	Training	
Manchester	56.15	47.24
Solihull	41.06	32.04

Figure 3.6   Extract from the Regionalisation Newsletter of the National Asylum Support Service (NASS). Source: Author's figure adapted from the National Asylum Support Service Regionalisation Project Newsletter, August 2005 issue, p. 4.

being the fastest at determining cases, completing the work in less than half the time allocated for standard cases, whereas the Manchester office was humiliated because it was missing both of the listed targets.

The scripting of asylum seekers' cases for welfare support in numeric and time-sensitive ways creates an artificial informational environment within which decision makers are under pressure to reach conclusions quickly. Asylum seekers are represented to NASS employees in simplistic, sanitised ways through this process. The distillation of a decision regarding an asylum seeker's claim for support into a minute-and-second count violently abstracts from the personal circumstances that underwrite each claim and threatens to allow the subjectivity, complexity and case-specificity of the practice of determining cases for welfare support to be overlooked.

'Find Your Local Asylum Support Team'[39]

The exclusivity and remoteness of the Portishead NASS office was not an isolated case. The reputation NASS accrued through the early 2000s for bureaucratic inaccessibility eventually led to its formal closure in 2006 and its replacement with so-called Regional Asylum Teams (RATs). Nevertheless, 'the central features of NASS – compulsory dispersal to slum housing and sub-subsistence support – remained untouched, surviving the abolition of NASS as a separate agency' (Webber, 2012, p. 96). Since then, despite numerous restructures, the dispersal of asylum seekers has continued while both '[t]he legislation in respect of eligibility for asylum support, and the categories of support available, has not changed' (Hansard, 28 January 2013). Indeed, the emphasis on RATs since the mid-2000s has increased the distance of immigration decision makers from their subjects and set them into even fiercer competition with each other over abstract metrics. In this section I demonstrate that the exclusivity of the Portishead NASS office was not an isolated example, and that the creation of RATs served to increase this exclusivity even further.

There were 12 NASS offices created in the early 2000s, each of which shifted the work of deciding cases for support away from city councils, situating back-office asylum work in more exclusive, whiter locations not just in Portishead – where the office was located in a postcode that is 95.1% white[40] – but around the United Kingdom. In Cardiff, for example, work was relocated from an area that was 66.5% white and had a population density 10% higher than the national average (that housed the council offices) to one that was 77% white and had a population density of just half the national average. In Glasgow and Manchester work was also relocated to areas of lower population density, and in Manchester, where the new NASS office was near to the airport, this also meant moving from an area with unemployment that was 9% higher than the national average to one with unemployment levels of just two-thirds of the national average. The census describes the first area as 'typically white or mixed race' and the second as simply 'typically white'.

Liverpool lost its asylum support casework to this new Manchester office. The shift from Liverpool to Manchester relocated the work of deciding claims for support to an area nearly 2000 places better off on the index of multiple deprivation (a composite scale of deprivation of around 32,000 places in total[41]) and from an area that is 70.2% white to one that is 79.8% white. A similar picture emerges in the cases of the Birmingham, Bradford, Hull and Nottingham NASS offices. In each case work was relocated to areas of lower population density and from areas with higher unemployment rates than the national average to areas with lower rates than average. Each move also involved relocation to an area that was substantially better off in terms of the index of multiple deprivation (by 7465, 8408, 8502 and 17,577 places respectively) as well as to whiter areas. In the case of Birmingham, for example, the move entailed relocation from an area that is 51.9% white to one that is 93.8% white, and the Bradford move entailed relocation from an area that is 21.1% white to one that is 72.1% white. Similarly in the case of Hull whiteness of the area surrounding the office increased from 51.0% to 72.1%, and in the Nottingham case from 28.5% to 93.9%.

In Newcastle the work was shifted from an area that is 53.8% white to one that is 69.3% white, and in Leeds the office was relocated from an area described by checkmyarea.com (a website for house buyers and local residents) as 'singles and couples in small terraced properties' to 'executive households in suburban terraces and semi-detached houses'. The Leeds move also involved shifting to an area with lower population density and from an area that is 45.7% white to one that is 72.1% white.

One of few exceptions was the abandonment of the work of determining cases for asylum support in Coventry, which shifted work from Coventry to the NASS office in Peterborough, which had higher population density and unemployment levels (although still higher percentage whiteness of 55% in comparison to 31.5% at the Coventry Council House). When the NASS offices were replaced by RATs, however, the work from the Peterborough office was shifted to Solihull, and to an area described as being populated by 'affluent, mature families and couples in large exclusive detached homes' by checkmyarea.com. At the Solihull office population density is just 27% of the national average, unemployment just 44% of average and percentage whiteness 82.9%. The shift from Peterborough to Solihull entailed a move to an area 23,247 places better off on the multiple deprivation index, thereby more than cancelling out the reduction in exclusivity associated with the creation of the NASS Peterborough office.

The new RAT offices, created in the mid-2000s, have a greater hinterland than either the city councils or the NASS offices had. Only six offices now serve the entire United Kingdom, including two international offices ('Wales and the South West' and 'Scotland and Northern Ireland'). When the RATs were created, offices remained in Cardiff, Glasgow, Manchester, Solihull, London and Newcastle, but disappeared from Leeds, Castle Donington in the East Midlands, Peterborough and Portishead. For those areas that lost an office, the physical remoteness of regional offices from the peripheries of their catchments is an obvious consequence of the reorganisation. The work that used to be carried out in Portishead, for example, is now carried out in Cardiff in Wales, more than 120 miles more distant from Bristol city centre than Portishead. This Welsh office also serves Plymouth, which is 151 miles away. What is more, the loss of these offices entailed further exclusiveness and social remoteness, continuing the tendency that was established when the regional NASS offices were created. The Manchester office, for example, which now serves the entire North-West, is in an area of population density lower than that of the former Leeds NASS office. Similarly the Solihull office that took over the work of the Castle Donington office has comparable population density, lower unemployment and an index of multiple deprivation rank over 6000 places better off than the area the Castle Donington office was in.

Overall, the creation of NASS in the early 2000s and the creation of RATs in the mid-2000s ensured the physical and social remoteness of decision makers from their subjects. Doubtless, Home Office managers would argue that there is no need for face to face, across-the-counter, contact between decision makers and asylum seekers. They might also point to the dwindling number of asylum claims to the United Kingdom (Figure 3.1), which necessitates fewer offices to deal with the paperwork associated with claims, in order to justify the greater hinterland of offices. But this does nothing to account for the steadily increasing social and racial remoteness of offices and exclusivity of office locations, which, whether intentional or not, entails the buffering and insulation of decision makers and prepares the way for adiaphoric competition between them over abstract metrics.

Conclusion

In response to Nikolas Rose's question of 'how [are]…bureaucrats and civil servants to be governed?' (Rose, 1999, p. 149), immigration sector employees' systematic estrangement from their subjects through processes of insulation, buffering and the nurturing of competition between them, constitutes an important part of the answer. Through these spatial innovations, NASS employees in the early 2000s were licensed to treat asylum seekers as if they did not raise questions of moral concern. Moral conundrums and the morally repugnant consequences of moving asylum seekers at short notice under the threat of destitution, housing them in inappropriate conditions, failing to provide sufficient financial support for them and allowing terrifying removal attempts to befall them were simply not allowed to confront the orchestrators or even the perpetrators of the system, who were instead allowed to concentrate on maximising highly abstracted metrics that bore very little resemblance to the actions themselves. Bauman (1989, p. 215) describes how 'Stretching the distance between action and its consequence beyond the reach of moral impulse [and] dissembling…human objects of action into aggregates of functionally specific traits, held separate so that the occasion for re-assembling the face does not arise' is the surest way to ensure that 'action can be free from moral evaluation' (Bauman, 1989, p. 215). This process in the case of NASS ultimately fell most heavily upon families, vulnerable asylum seekers and the most obliging, because it is these that offered the fastest and cheapest way to meet deportation targets.

These observations allow us to enrich our understanding of how moral distancing works through a bureaucracy. Not only are subjects distanced from decision makers, but decision makers are also distanced from subjects. In both Chapter Two and in this chapter, however, I have generally focused on literal forms of distancing that separate decision makers and subjects. As theorists of moral distance have shown, this literal separation is connected to moral distance, which is a particular type of moral estrangement. In the case of NASS there is a clear correlation between physical distance and moral distance, because it was through the process of physically segregating decision makers from the everyday life of asylum seekers that moral estrangement and indifference was achieved. In the next chapter, however, I will examine a different situation in which moral estrangement occurs despite physical closeness. As it turns out, the tendency of bureaucratic forms to create indifference among decision makers is by no means cancelled out by physical closeness between decision makers and subjects. Moral distance, in other words, is only one mechanism through which estrangement and indifference are nurtured.

Notes

1 Interview with immigration solicitor, Bristol, 3 November 2005.
2 Although the interviewees at the NASS office mostly refused to be recorded, quotes are based upon scratch notes made during the interview and written up into a full account shortly after the interviews finished.
3 Interview with NASS employee, 5 May 2006.
4 The British National Party (BNP) is a nationalist political party. The *Daily Mail* newspaper published various explicitly anti-asylum stories during the 2000s, to the extent that the BNP and the *Daily Mail* became by-words for an anti-immigrant perspective among refugee support workers and activists during the mid-2000s.
5 Interview with NASS employee, 5 May 2006.
6 Interview with NASS employee, May 2005.
7 Matthew Hull (2012) has discussed the issue of the centrality of paper to modern governments, while Jonathon Darling (2014) has examined the importance of letters in asylum seekers' relationships with the Home Office.
8 See Gill (2014) for a fuller discussion.
9 Interview with volunteer coordinator, Bristol, June 2006.
10 This quotation is taken from the Memorandum Submitted by Citizens Advice to the House of Commons Home Affairs Select Committee who announced an inquiry into asylum applications on 25 February 2003.
11 Interview with volunteer, Bristol, 7 November 2005.
12 Interview with local government employee, Gloucester, 20 June 2006.
13 Interview with NASS employee, 5 May 2006.
14 Interview with NASS employee, May 2006.
15 Interview with NASS employee, June 2006.
16 Interview with NASS employee, May 2006.
17 Interview with NASS employee, Portishead, June 2006.
18 Interview with NASS employee, Portishead, May 2006.
19 Interview with NASS employee, Portishead, June 2006.
20 Interview with NASS employee, Portishead, May 2006.
21 Interview with charity worker, Bristol, June 2006.
22 Interview with resident of Portishead, 7 June 2006.
23 Interview with church leader and resident of Portishead, June 2006.
24 Interview with resident of Portishead, 7 June 2006.
25 Interview with church leader and resident of Portishead, June 2006.
26 Interview with resident of Portishead, June 2006.

27 Interview with resident of Portishead, June 2006.
28 Interview with NASS employee, Portishead, May 2006.
29 Interview with NASS employee, Portishead, June 2006.
30 Interview with charity worker, Bristol, June 2006.
31 Interview with activist, Bristol, 13 November 2005.
32 Interview with local government employee, Gloucester, 20 June 2006.
33 Interview with volunteer refugee sector co-ordinator, Bristol, 17 May and 7 June 2006.
34 Interview with NASS employee, Portishead, June 2006.
35 Interview with volunteer refugee sector co-ordinator, Bristol, 17 May and 7 June 2006.
36 Interview with NASS employee, Portishead, June 2006.
37 Often asylum seekers are required to regularly visit reporting centres, often located in local police stations, in order to indicate to authorities that they have not absconded.
38 There was a fast-track procedure for claimants coming from so-called 'White List' countries, that is countries that had been deemed to be generally safe by the Home Office, meaning that claims for asylum from these countries were viewed as manifestly unfounded and claimants were entitled to only a pared down version of the asylum determination procedure.
39 The subheading 'Find your local asylum support team' is taken from the Home Office website that gives details of asylum support (Home Office, 2014b). The figures quoted in this section are the author's calculations based on data from the Department for Communities and Local Government (2014) for the multiple deprivation index, the Office for National Statistics (2014) for ethnicity, unemployment and population density, and check-myarea.com for descriptions of local areas. I acknowledge valuable assistance from Abigail Grace for this section.
40 Ethnicity data on percentage whiteness in this section is derived from British census data and made available by the Office for National Statistics at the following web address http://www.nomisweb.co.uk/census/2011/ks201ew (accessed 22 July 2015). The percentage given refers to amount of people who are 'White: English/Welsh/Scottish/Northern Irish/British' not the broader category of 'white' in terms of visibly white as this includes certain groups that can be classed as minorities such as 'gypsy or Irish traveller' and 'other white', which may include white people from non-UK origins. This information was not available for Scotland. Where a specific postcode area is not available from the dataset, an adjacent postcode area is used. This was found using Google Maps and Postcode Finder to find the closest possible postcode area to the building housing the offices under discussion (no further than 1/4 mile).
41 The Index of Multiple Deprivation is an index that ranks 32,482 small areas of the United Kingdom by income, employment, health and disability, education, skills and training, barriers to housing and other services, crime and living environment. The area ranked '1' has highest deprivation and the area ranked '32,482' lowest deprivation.

Chapter Four
Distance at Close Quarters

It has sometimes been held that merely by assembling people without regard for race, color, religion, or national origin, we can thereby destroy stereotypes and develop friendly attitudes. The case is not so simple.

Gordon W. Allport, 1954, p. 261

I am standing outside Lunar House in Croydon, London, the headquarters of immigration control in the United Kingdom in July 2013 in the baking sun.[1] The office has closed its front desk today in anticipation of trouble. The English Volunteer Force (EVF), a right-wing, nationalist political lobby group, having recently been banned from marching on mosques in London, have chosen Lunar House as their next target. They are convinced that immigration controls are too lax in the United Kingdom and are planning to vent their anger at staff and immigration policy on the small concrete plaza outside the main Lunar House complex. Unite Against Fascism (UAF), an anti-racist pressure group that promotes diversity in British communities, has joined forces with the Public and Commercial Services (PCS) Union – the union that represents many of the frontline staff in Lunar House – to stage a counter-protest against the EVF. The composition of the counter-protest is striking. There are middle-managers employed by the Home Office in immigration control here, standing next to frontline immigration officers, standing next to radical anti-fascist activists and

Nothing Personal?: Geographies of Governing and Activism in the British Asylum System, First Edition. Nick Gill.

Figure 4.1   Anti-fascist protest outside Lunar House. Author's photograph, 2013.

refugees. An anti-fascist flag flies proudly outside immigration HQ as if there was nothing unusual about it at all (see Figure 4.1).

The English Volunteer Force finally arrive, daubed in white and red face paint, flying Union Jacks and chanting nationalistic slogans. For over an hour the counter-protest, which easily outnumbers the EVF, are united in their chanting in reply. 'Whose streets?' cries a protestor with a megaphone; 'Our streets!' we thunder in response. The EVF marches peacefully past the waiting counter-protest and although a small scuffle occurs when someone tries to steal one of the EVF flags, the EVF members have been well briefed. They have realised that violence and thuggery do nothing to improve their public image and they do not rise to the provocation. Instead they take up a position in a small, cordoned area opposite the counter-protest and begin to exchange chants and insults.

The employees of Lunar House are huddled towards the back of the anti-fascist protest, joining in with some of the chants but looking decidedly uncomfortable about others. At one point a No Borders group takes the megaphone and belts out the slogan 'No Borders, No Nations, Stop the

Deportations'. As they do, I watch the immigration officials turn away and chat amongst themselves. As the chanting wears on through half an hour, and then an hour, the officials fall silent or talk to each other rather than to the other anti-fascists. I talk to one of them, a trade union member and an immigration officer employed in Lunar House who was disgusted with the fact that the management of Lunar House had done nothing to prevent the EVF march. 'The workforce in Lunar House is 65% BME [British Minority Ethnic]', he explained,[2] 'and many of our members could be intimidated by the EVF march'. It's clear that his motivation for attending the anti-fascist counter-protest is laudable, but very different from that of many of the other attendees. Although many of the protesters oppose the current level of immigration restrictions and, by implication, the very rationale for the work that Lunar House employees carry out, he limits his concern to the safety of his colleagues whilst executing the work of border control and enforcement.

In a similar vein, another immigration officer and union member explained his frustration at what he saw as the negligence of Lunar House management in protecting staff during the EVF march. The march was staged, with the help of heavy police presence, on a forecourt at the front of the Lunar House building, and the management at Lunar House had announced that this forecourt constituted part of the public street and not part of their own premises (see Figure 4.2). 'This is despite disciplining an employee recently for fighting on the forecourt, which implies that the forecourt *is* Lunar House's concern', explained the officer. 'Saying that the courtyard is not Lunar House land has meant that the senior management team are just keeping their heads down. They want this [march] to pass by. As far as they are concerned it's not their problem.'

According to the same officer, the management team had also issued warnings to the Lunar House workforce that political participation in the protest or counter-protest could lead to disciplinary action. For this officer, the denial and indifference of the senior management team to the EVF march, and their confrontational stance to the political activities of their workforce, typified the general managerial approach at Lunar House that was partly responsible for his presence at the protest. He was frustrated by the threat of job losses too, and the fact that 'they have just doubled targets so everyone is supposed to work twice as hard'.

As the EVF marches on towards other sites in central London on their itinerary, the counter-protest begins to wind down. The immigration officers are among the first to leave, politely nodding to the organisers but keeping a good distance from the rest of the protesters. So while the counter-protest provided brief contact between migrants and immigration enforcement officers it was over within a couple of hours and any sense of unity that the chanting and common adversary of the EVF had generated quickly dispersed. The officers face very different constraints and have

Figure 4.2   Forecourt outside Lunar House and site of the English Volunteer Force (EVF) protest and counter-protest. Author's photograph, 2013.

very different priorities to the migrants and migrant support groups that they have just stood next to. Although they have come into contact with each other, their differences and the fleetingness of their meeting have precluded an encounter from occurring even in close proximity.

In this chapter I examine how encounters and meaningful interaction are averted at close quarters. While much of the work of British border control can be routinised and carried out at a distance via telecommunications, such as the work of NASS that I discussed in the previous chapter, there are occasions during the processing of an asylum application when an official must meet a migrant who is claiming asylum. When a claim is first lodged there is such a wealth of information that needs to be collected that an interview is the most efficient approach, and a second interview is usually used to explore the claims of applicants in more detail. And, should a claim be refused, asylum applicants have a legal right to a tribunal hearing during which they, as well as a judge and a representative of the Home Office, usually attend in person. Although both the asylum interview and the appeal are face to face contact events, they are carefully choreographed, so that 'who gets close to whom, and under what circumstances, is not left to chance' (Fortier, 2007, p. 104). It transpires that there are various ways in which the sort of morally impactful interaction that jolted Mr Brimelow to petition his superiors on behalf of Nataliya (see Chapter Two) might be avoided even when an applicant is sitting right across the desk. As I will show, it is possible, and in many cases routine, to be brought face to face with another and remain indifferent towards them.

This argument entails talking about 'encounters' in a different way to Levinas. There is a significant literature that examines prejudice, meaningful interaction and contact within and beyond geography, and throughout this literature meaningful contact is generally understood as entailing interaction 'that actually changes values and translates beyond the specifics of the individual moment' (Valentine, 2008, p. 325). Commentators have disagreed over precisely what meaningful contact or interactions involve, with some emphasising their results (e.g. 'new influences and new friendships', Amin, 2002a, p. 970), others emphasising their ability to 'shift consciousness' (Wilson, 2013, p. 76) and others foregrounding the degree to which they involve 'deep' as opposed to 'surface' level emotional responses (Hemming, 2011). The majority of these discussions, however, do not employ an understanding of encounters in terms of an intense, non-synchronous and non-coincident experience of difference like Levinas does (see Barnett, 2005). This is important because, basing our approach on Levinas, we might argue that an encounter precedes interaction, and even, given that an encounter is an experience of rupture, that interaction signals a degree of commonality and reciprocity that we might associate with equality and similarity, rather than difference, and hence that where meaningful interaction occurs encounters cannot.

For the purposes of my argument here, however, I take a more pragmatic approach that employs a rather less demanding definition of encounters. I refer to encounters as morally obligating interactions (Reader, 2003). That is not to say that I understand meaningfulness and morally obligating interaction to be equivalent: meaningfulness is subjective and could conceivably (be felt to) arise without experiencing difference with a partner in an interaction, and certainly without becoming concerned for their welfare or morally obligated to them. It is easy to imagine meaningful interaction between a teacher and a pupil, for instance, without either of them experiencing new moral demands as a result. Nevertheless, it is reasonable to posit that morally obligating encounters are meaningful to those that experience them, so the discussion of meaningful interaction that follows bears upon the notion of the encounter in so far as morally obligating encounters constitute a certain, specific, sort of meaningful contact.

My argument is that we need to give far more attention to how situations in which morally obligating encounters might occur are transformed into situations in which impersonal, unconstructive, negative or meaningless interaction takes place. There is much mileage in examining the features of organisations that ensure that the conditions for meaningful contact are violated and that a meeting or contact event does not become a morally demanding encounter. More broadly, the chapter demonstrates that moral estrangement is by no means the preserve of situations in which decision makers and subjects are physically remote from each other. If anything, the bureaucratic tendency towards moral estrangement reaches a crescendo at close quarters, and consequently produces its most sophisticated and disturbing distancing techniques in situations where decision makers and subjects meet.

The Conditions of Contact

In the United Kingdom, the Home Office is responsible for processing applications for asylum. Applicants are required first to attend an interview at which they are asked for biographical and biometric data, details on their travel history, and basic information about why they are seeking protection. This is followed by a second, more substantial asylum interview with a Home Office case worker, and ideally (although all too infrequently) in the presence of a legal representative, which is designed to elicit the basis of the claim in some detail. In roughly three-quarters of cases, the Home Office decision is a refusal of asylum, which usually means that the applicant has no legal right to remain in the United Kingdom. Asylum applicants generally have a right of appeal against this decision to an independent Tribunal known as the First Tier Tribunal (Immigration and Asylum Chamber) (hereafter referred to as 'the Tribunal'). Each substantive appeal[3] is heard before an immigration judge, and usually involves the applicant, their legal representative, the respondent (who is a Home Office representative), an interpreter if required, witnesses if called, and Tribunal clerks (who are 'silent' actors who assist in the smooth running of the process). After the hearing, the immigration judge must produce a written determination, either allowing or dismissing the appeal. In this chapter I examine these two events – the asylum interview (both screening interviews and substantive interviews) and the appeal hearing – in order to determine how their choreographing impacts upon the likelihood of morally demanding encounters between applicants and decision makers.

As a first step towards grasping the different forms of contact that bureaucracies can nurture, it is vitally important to depart from the idea that contact between different groups will reliably be positive and friendly and lead to reduced antipathy. Although this notion has rightly been dismissed as 'starkly naïve' (Jahoda, 1987, p. 275), 'the adoption of the contact thesis outside of social psychology, and especially in policy circles, has been too uncritical [by lending too much credence] to sweeping, unequivocally optimistic conclusions of the dominant strand of contact scholarship about the positive effects of contact' (Matejskova and Leitner, 2011, p. 720).

In reality, there is no guarantee that 'cultural difference will somehow be dissolved by a process of mixing or hybridization of culture in public space' (Valentine, 2008, p. 324), especially outside the clinical experiments, most often conducted with college students, that contact researchers tend to generalise from (Dixon *et al.*, 2005, p. 706).

Meaningful contact and interaction can lead to a fresh understanding of, empathy towards and even intimacy with a previously othered subject (McLaren, 2003). It involves a 'something between' two people that links them together (Reader, 2003, p. 370). It involves 'something real...the inter-twinement of a bit of life...potentially generating moral obligations' (Reader, 2003, p. 372). These are demanding characteristics that jar with our natural inertia towards newness and our impulse to 'not welcome difference, trans-formation, and change' (LaVan, 2003, p. 6). Many contact situations therefore leave values 'unmoved, and even hardened' (Valentine, 2008, p. 325). As Allport (1954, p. 263) observed, 'Where segregation is the custom contacts are casual, or else firmly frozen into superordinate-subordinate relationships. ... Such contact does *not* dispel prejudice; it seems more likely to increase it. ... The more contact the more trouble'. So although certain types of contact offer the 'possibility for change' (Leitner, 2012, p. 840), and the 'potential...to disrupt' (Leitner, 2012, p. 840), other forms of contact work in the opposite direction, sometimes sparking fear and anxiety of others and sometimes simply leaving pre-existing attitudes unchanged.

From the perspective of an encounter-averse bureaucratic administration concerned about the unpredictable risks associated with personal interaction between its subjects and its employees, this insight can be seen as something of a reprieve. As it turns out, there are various features of the institutional arrangement of asylum interviews and appeals that ensure that these events generally fall short of morally demanding encounters. So while researchers have investigated 'the spaces of interaction that may enable meaningful encounters' (Valentine, 2008, p. 325; see also Askins and Pain, 2011, p. 803) my interest is in how interactions have been choreographed in order to *disable* meaningful interaction, nurture negative, empty and meaningless contact and thereby rule out encounters.

Allport (1954) set out a series of conditions for contact to be meaningful and prejudice-reducing, stressing that 'the effect of contact will depend upon the kind of association that occurs' (Allport, 1954, p. 262). These con-ditions include, first, the frequency and duration of contact. Contact that is fleeting, Allport observed, is less likely to lead to prejudice reduction than contact that is sustained and/or repeated.[4] Second, he stressed what he called the 'role aspects of contact' (Allport, 1954, p. 262) including whether 'the relationship [is] one of competitive or cooperative activity' (Allport, 1954, p. 262). Where the latter case is true and the two parties understand themselves to be 'in pursuit of common goals' (Allport, 1954, p. 281), Allport argued that the likelihood of meaningful contact is far greater. 'Only the type of contact that leads people to *do* things together is likely to result in changed attitudes' (Allport, 1954, p. 276), he stated. Third, interaction is likely to reduce prejudice when the 'contact is sanctioned by institutional supports [such as] law [or] custom' (Allport, 1954, p. 281). Fourth, Allport identified the 'social atmosphere surrounding the contact' (Allport, 1954, p. 262) including whether the contact feels real or artificial and whether it is regarded as 'important and intimate or as trivial and transient' (Allport, 1954, p. 262). Given that interaction needs to be relaxed and free-flowing for relationships to develop or for prejudice reduction to occur, stilted, highly formal and prescribed forms of interaction are likely to be counterproductive, especially if the subject lacks 'basic security in his own life, [is] fearful and suspicious' (Allport, 1954, p. 263), or has had negative previous experience of the dominant group in question. Fifth, Allport stressed the importance of the 'status aspects of contact' (Allport, 1954, p. 262) during the interaction: when equal status contact between groups can be achieved this is much more conducive to prejudice reduction. In summary, '[o]ptimal contact [requires that] the contact is sanctioned by relevant authorities, it is cooperative, people engaged in contact are working toward common goals, and the people engaged in contact have equal status' (Barlow *et al.*, 2012, p. 1630). Where these conditions are not met, contact theorists have demonstrated that contact is far less likely to result in meaningful interaction and reduced prejudice and may very well act in the opposite direction (Pettigrew and Tropp, 2006).

Inside Lunar House

Lunar House is a 20-storey office block in the centre of Croydon, a busy London suburb (see Figure 4.3). It houses the headquarters of British immigration control and is the place where people can apply for British citizenship or claim asylum if they did not do so upon entering the United Kingdom. In 2005 there were just under 2000 staff based in Lunar House with a further 4400 in neighbouring towers and offices around Croydon (Back *et al.*, 2005). Staff are involved in a wide range of aspects of immigration application determination, including asylum application management and processing as well as fraud detection, security and enforcement functions. This section describes the work that is carried out at Lunar House with an emphasis on its administrative inefficiency, the lack of adequate training for staff, and the stress and pressures that staff experience at the site. This sets the scene for the following section that assesses the degree to which Allport's conditions for meaningful interaction are met during the asylum interviews conducted there.

My research participants consistently recalled the crimes of bureaucracy committed at Lunar House that give the impression of a beleaguered and

Figure 4.3 Lunar House with sign and Union Jack. Author's photograph, 2013.

woefully under-resourced system. Documents like identity cards and pass-
ports are frequently mislaid, for example. 'I was just flabbergasted at the
complete ineptness of the situation', recalled one activist who had helped an
asylum seeker whose identity card had been mislaid:

> I mean I spent five hours on the telephone retelling the story every single time
> only to be told 'well, we can't deal with it, we'll have to pass you on', and we
> went round and round and round in circles and ended up back at the reception
> desk! The only person whose name they would give me was the person in
> charge of the whole immigration service and the Minister of State. It's
> Orwellian.[5]

Similarly, a former asylum seeker who worked to help others navigate the
application process during the 2000s asserted that:

> Communication is always poor in Lunar House. We send lots of people different
> things: original passports, I.D. cards, affidavits for their case. When they
> received them they lost them. And people get angry with me or my colleagues

and say how it is my fault. No it wasn't my fault. We sent it by recorded delivery. They didn't put it into the right system. And then they said 'well it is in the system, somewhere'. They didn't tell us where.[6]

So commonplace were the bureaucratic hiccups, and so costly the human consequences, that South London Citizens (SLC), a collection of schools, churches, mosques, synagogues, charities and residents' groups committed to taking action for the common good of the people of South London, commissioned an investigation and report into the shortcomings and inadequacies of Lunar House in 2005, entitled *A Humane Service for Global Citizens* (Back *et al.*, 2005). One of the contributors described the Immigration and Nationality Directorate (IND) as 'impenetrable…it's a system that puts rules before people…there is a huge discrepancy between what they say they are doing and what happens at the routine, everyday level…it's a faceless bureaucracy which can crush people because it has such a definitive role in life or death.'[7] Even a senior manager at Lunar House admitted that 'we do have communication problems in that it's difficult to track people down internally. I mean we've got 17,000 people in the IND [but] if you want to search for a department you can't do it.'[8]

These deficiencies resulted in unacceptably long delays in organising asylum interviews and making or communicating decisions. 'My understanding of how things work', one asylum support worker who had visited Lunar House numerous times noted,

is that as soon as a new piece of legislation comes into force the people who are still under the previous pieces of legislation end up going towards the back of the queues because the new legislation gets implemented first. Staff are retrained, new staff members are taken on to deal with the latest legislation and that means that there is a real problem then for people who are still waiting for answers or had their last decision before this came into force. They have to wait and find out what happens next and it will just take longer because the teams dealing with the previous regimes are much smaller. Now, shockingly, we have still got people who came in before 1993 and haven't had a first decision.

Interview, asylum support worker, 2006[9]

The effect of pointless waiting was also felt on a daily basis. One asylum seeker fleeing the threat of persecution for his political beliefs and activities as a journalist in an African country described his first experience of British bureaucracy at the Lunar House complex, just one day after his arrival in the country, which resulted in him needlessly sleeping rough:

I took the train and I got out and I remember the name it was Piccadilly. I found someone who could speak French and then he helped me to go somewhere 'cos it was late. And he explained how to take a bus and how to buy a ticket and then I went there [to Lunar House]. I get inside and then they tell me it is too late to come so you have to wait for tomorrow. So I don't have place to go, so I stayed there just walking, walking around until morning. [It was] very freezing, it was very freezing at this time; it was March, very freezing. In the morning they opened. When they opened I was the first in the queue and I get inside and they told me it's not this place you have to go to another place.

Interview, former asylum seeker, 2006

One former asylum seeker noted the lack of 'even basic standards of care, be it providing seating for the elderly standing for hours in the cold, or the omission of a ticket queuing system when you finally entered the building',[10] which for them and many other activists familiar with the site, pointed towards a flawed perception of how to treat people in a reasonable manner. As one activist noted wryly:

> To speak about 'clients' and so on, and 'managed' migration, and all of those seemingly anodyne and banal ways of talking about a simple process, well how that compares to what it means to go through those gates, those doors, to go inside the offices of immigration and to be subject to the kind of routine humiliations that that involves, you know it just doesn't stack up. I can't think of any business that would treat its customers like that.[11]

The conditions of work and the lack of training of the workforce undoubtedly compounded the difficulties. During the 2000s each new caseworker received only 11 days of initial training followed by a minimum of 11 days of mentoring before they were allowed to make independent decisions on claims for asylum (compared to six months training in Germany for caseworkers who had no previous legal experience). Before February 2004 caseworkers also did not require A-levels.[12] For workers in Croydon, they consequently began to view the employment opportunities in Lunar House as a stop-gap or stepping-stone to better things rather than a career. 'It was just a job really, it wasn't a passion of ours', a former intelligence officer, employed to investigate instances of abuse and benefit fraud among asylum seekers during the 2000s, recalled.[13] We got hardly any training' she continued, 'we just went straight in. Straight in. We had a test that…had absolutely nothing to do with the job we ending up doing, but they kind of just throw you in at the deep end and then just kind of leave you there.' In the end, this sense of dissatisfaction contributed to the same officer's decision to leave her job, because she felt that she was underemployed, having completed a degree before she took the position. 'My [colleague] didn't have a degree', she recalled, 'and I don't think she's got A-levels either and that was another reason I left the job because I thought "well I could have just come here at sixteen", you know? Instead of going to university and being there for however many years.'

Alongside the lack of training, other forms of support were also lacking. The PCS union that represents the workforce in Lunar House drew attention to the lack of progression opportunities, the high turnover of staff, low morale, lack of faith in management, and the constant fear of media scrutiny and criticism. One activist also expressed concern over the lack of awareness of secondary trauma,[14] an issue to which we return in Chapter Five:

> I mean there are real issues to do with secondary trauma for staff members. The way people cope with hearing stories that are really quite difficult to hear, if they're not getting adequate support themselves, is to burn out or else shut down. It feeds into cultural disbelief ultimately, particularly when you've got very young staff like school leavers. There is not sufficient support of staff members for them to deal with any traumatic feelings that they have, their own traumatisation on a secondary level from interviewing people, from reading the documents, from whatever. And it gets to a stage for some people that the only way that they can deal with this is to shut down and say 'this can't be happening, they're all liars'.
>
> Interview, asylum support worker, London, 2006[15]

The targets, the lack of support and the threat of negative media attention meant that staff retention and stress were recurring problems. 'We're either too easy on asylum seekers or we're rightwing fascists', declared the PCS Union branch secretary in 2003, capturing his belief that staff were in a 'no-win situation'.[16] The SLC report conveyed similar concerns. 'Half the workers...had experienced stress-related health problems within the last year', the report revealed (Back *et al.*, 2005, p. 62). 'Causes of stress [included] the difficult nature of the work, the volume of work, unrealistic targets, long hours on the front line, high pressure caused by a hostile media climate, case-fatigue, frequent policy changes, threat of job cuts, poor management [and] lack of support' (Back *et al.*, 2005, p. 62). Workers quoted in the report admitted that they felt 'anxious, frustrated and demotivated' (Back *et al.*, 2005, p. 63). 'I am disappointed in myself', one decision maker confessed, 'because I end up acting in an uncaring and unsupporting way when dealing with customers' (Back *et al.*, 2005, p. 63) while another conceded that '[i]t is very difficult not to become hardened when dealing with people due to the lack of leeway and support the government offers its employees' (Back *et al.*, 2005, p. 63). 'We're trained to do the work', another observed, 'but not at this volume' (Back *et al.*, 2005, p. 65).

Staff turnover was, unsurprisingly, high, especially among more experienced staff. In 2006 34% of staff leavers from the IND had between three and five years' experience in their roles, meaning that they took substantial knowledge and experience with them when they left, and by 2013 concerns around staff retention were still rife. In a members' briefing in 2013, the PCS union reported that many of the same problems endured, including the constantly 'increasing targets' and managerial tendency to demand 'more for less'. They described 'a continued erosion of staff numbers as the rush to the door continues unabated'.

'Although it was boring, it was stressful', the former intelligence officer recalled in reflecting upon how difficult conditions became before she left her post:

> I used to leave there nearly every day with a migraine. I used to spend a lot of time in front of the computer just filing and just trying to sort out people into different categories to investigate at a later date and...I would have targets I would have to get through: however many in a week. And sometimes that was virtually impossible because phoning people all the time, you know, people don't like cold callers...I couldn't wait to get out of there...I remember saying to [a relative] 'I can't take any more'.[17]

Overlaying all of these issues, the spectre of racism among the workforce was often in attendance. Many of the workers at Lunar House lived in and around Croydon, which suffered chronic shortages of affordable housing and council housing during the 2000s. 'They [meaning asylum seekers in Croydon] would get first for the housing', the same officer recalled with a tinge of bitterness. 'So somebody could have lived here all their life, worked here, contributed, and then they come and they get, you know, the newest accommodation...it's not fair'. From this perspective she described her colleagues' attitudes – including caseworkers, interviewers and other enforcement personnel – to the asylum seekers in their charge:

> I know sometimes the discussions that people would have with me in the office I just would excuse myself from. ... You might hear people saying things as you walk past them and they say 'Oh, no offence',[18] little things like that. I think the longer you worked there you become less open minded, the more you start to think the rest of the people shouldn't be here, they should go back. I never really heard anything positive about [applicants] while I was there. [Staff] would never be outwardly racist, they would be quite careful about what they said, but the undertone was always there.[19]

Asylum Interviews

Asylum seekers are pitched into this working environment in the hope of fair, consistent and well-considered treatment at a pivotal moment in their lives when they may be traumatised and disorientated. 'The significance of the asylum interview for the decision making process cannot be overstated', Crawley (2010, p. 163) asserts, while Woolley (2014) points out that the account that asylum seekers give of themselves and their experiences at interview can be the difference between life and death. Concerns have been raised, however, about the way in which evidence gathered at asylum interviews has been used against applicants at a later date. For example, whereas inconsistencies between accounts given during either the screening or substantive asylum interview and later in the application process are routinely used to undermine the credibility of applicants, psychologists have established that inconsistency is not only common when two descriptions are given of the same event at different times, but that inconsistencies in peripheral details that are not central to the gist of the account are especially likely if the events recalled are traumatic ones (Herlihy and Turner, 2006). Moreover, in the case of female applicants' interviews 'asylum decision makers often [invoke] factors such as delayed disclosure…and a calm or overly emotional demeanour on the part of the female applicant to justify suspicion regarding allegations of rape' (Memon, 2012, p. 678). This tactic betrays a lack of understanding of 'the different ways in which a witness may respond to trauma and [its] complex effects on testimony' (Memon, 2012, p. 678), including the effects of 'shame, dissociation and psychopathology in disclosure' (Bögner et al., 2007, p. 75). More generally, given that 'a Home Office interview can be a stressful and anxiety-provoking event' (Bögner et al., 2007, p. 75), psychologists have warned that the 'circumstances of the interview process itself' can profoundly interfere with disclosure (Bögner et al., 2007, p. 75).

I now turn to an assessment of how well Allport's conditions for meaningful contact are met during screening and substantive asylum interviews such as those held at Lunar House in the mid-2000s.[20] Only one of his five conditions is not obviously violated: the meeting between interviewer and interviewee is legally legitimate, meaning that the meeting is 'sanctioned by institutional supports' in the way Allport (1954, p. 281) advocates. A second of Allport's conditions – that the two parties pursue common goals – is arguable. Senior managers at Lunar House were extremely keen to emphasise the goals that they shared with asylum applicants and activists. Upon hearing about the SLC report, for example, management welcomed the activists' efforts ('it's good to have a bit of external pressure actually', a senior manager confided to me[21]) and expressed their eagerness to cooperate with the initiative. 'The reason why we want to cooperate with them is that their objective is exactly the same as our objective', the same manager continued. 'You know, let's make this better, we want to improve customer service, we want to make things good for staff…We're always looking to improve our service'. In particular, the manager emphasised the common desire to not have applicants waiting too long inside or outside the building before their interviews occur. 'From our point of view', she explained, 'we've got a lot of cost constraints so we don't want to have lots of people waiting [because of] the added security and the added buildings.'

Although the manager did not admit it, the embarrassment that long queues outside Lunar House caused the IND and the government was probably another factor in wanting to work with SLC to reduce waiting times. Lunar House became notorious for long queues during the 2000s. 'It is clear from the sea of humanity that descends on Croydon each day that even 20 storeys of bureaucrats cannot cope with the workload', *The Observer* newspaper taunted in 2003 (Bright, 2003), for example, while the then Minister for Immigration, Tony McNulty, was forced to concede, during an interview on a high profile radio documentary in 2005, that the 'pig pen'-style queues were not satisfactory (BBC, 2005b).

So pronounced was the eagerness to cooperate with the activists' report and recommendations that the SLC activists themselves found the managers' attitude disorientating. 'In some ways you feel almost lost for words', one contributor explained,

> because it feels like the government have taken all those words which you are using to try and open up things or to argue about injustices or to argue about hypocrisy in the system and you hear the same words coming out of the mouths of senior Civil Servants and senior politicians who have interests that seem very different.[22]

On closer inspection though, the interests of managers and applicants only coincided around very specific issues. The same senior manager explained that she preferred to try to reduce waiting times than to take any measures to make the waiting areas more comfortable. '[The activists] were saying "Oh, let's make things a lot more comfortable"', she recalled. 'Well actually it's not that comfortable, it's hard chairs...it doesn't matter if it's a hard seat because we only want them to be there for twenty minutes.' Her concerns to reduce waiting times were also related to the perceived need to deter 'people who try to abuse the system'. 'I mean the Home Office motto is a safe, just and tolerant society', she explained,

> so in some ways the more successful we are in making it safer, juster and more tolerant the more people will want to come over, so it's slightly ironic. The better we are in our objectives, the more people. But in terms of the people who try to abuse the system, we've put a lot more controls and checks and that acts as a deterrent and not only does it stop people today it actually stops people tomorrow because it's a big community out there, word gets around if it's more difficult to get in. If we do their case quicker and then actually remove them when they're not eligible to stay, then they think 'Is there any point in going to Britain, you're only there six weeks, you're locked into detention and then you're sent out?' So that's what we've got to do: process things quicker, faster, better decisions, quicker decisions, more removals. Then it acts as a deterrent.[23]

So although the management may see no conflict between their desire for faster processing and the desires of applicants, their wider concern to deter future fraudulent applications by processing claims faster indicates a very different rationale for wanting to speed up decision making. This conflict comes to a head when the desire to speed things up begins to undermine the quality of decision making itself: applicants do not want faster processing at the expense of well-considered decisions. Yet management's drive to reduce waiting times has resulted in sustained pressure on immigration staff to conduct interviews and reach decisions more and more hastily. The PCS union described planned increases in the target number of interview and decision completions in 2013 as 'eye watering' (Public and Commercial Services Union, 2013). 'If we want a decent Asylum system that deals sympathetically with victims of torture and persecution we must have a workforce fully trained and with time to interview applicants and time to come to well-reasoned decisions' the Union declared (Public and Commercial Services Union, 2013). If quality decision making is jeopardised by the apparently common goal of speeding up decisions then this is not a common goal at all, violating Allport's stipulation that groups work together with a common purpose.

The remaining three of Allport's conditions – concerning the frequency of the contact, the social atmosphere of the contact, and the status of the parties – are unequivocally violated. Although interviews are by no means fleeting, they do not constitute sustained contact in the way Allport prescribes. So although substantive interviews often last three or more hours, and constitute a gruelling ordeal for applicants as a result, they are one-off contact events that are hardly conducive to the development of 'acquaintance' (Allport, 1954, p. 268). In many cases applicants have waited a long time to be interviewed. One Kurdish refugee, who eventually won refugee status on appeal after eight years of waiting, described his feelings ahead of his substantive asylum interview at Lunar House, which came two and a half years after he initially claimed asylum at the border. 'I was very scared', he recalled,[24] 'because I didn't believe it. It was so long, at that time, after they told me six months and then it was two and a half years, it was awkward and weird.'

In terms of the social atmosphere of the contact '[t]he purpose of the [substantive] asylum interview', the Home Office writes, 'is to obtain details about why the applicant has made an application for asylum and/or leave to remain on human rights grounds. It is an opportunity for the interviewing officer to find out more about the applicant's fear of return to their country of nationality, and an opportunity for the applicant to elaborate on the background to his claim and introduce additional information' (UK Border Agency, 2007). As such, an interview is intended to be a fact-finding exercise rather than a conversation, during which the interviewer is required to give no information about him or herself, creating a distinct asymmetry in communication because applicants are expected to reveal some of the most intimate and difficult details of their own lives. This asymmetry does not provide the grounds for relaxed, free-flowing interaction.

In 2013 the Home Affairs Select Committee[25] conducted an inquiry into the asylum system in the United Kingdom and received over 100 written submissions from migrant support groups, activists and charities about the asylum determination process (Home Affairs Select Committee, 2013a, 2013b). The submissions highlighted a veritable catalogue of issues surrounding the social atmosphere of both substantive and screening asylum interviews, including the inappropriateness of the setting for the interview (which was described as 'hostile and intimidating') and the inappropriateness of the questioning techniques employed. 'Asylum seekers are not routinely told that they can choose to have an interviewer and an interpreter of their own gender',[26] one submission pointed out, while another noted that 'although asylum applicants are allowed to have legal representation in their critical first interview [meaning their screening interview at Lunar House] it isn't a requirement that they are advised of this (and many aren't).'[27] Others drew attention to the lack of privacy of the screening interview, and the lack of play facilities meaning that some asylum seekers had their children with them during their interviews. 'When I had my substantive interview' one former asylum seeker recalled,

> there were 5 people (myself, interpreter, my child and 2 staff from UKBA) in a small room. Every 5 mins, my daughter was interrupting asking for a drink or to go to the toilet or for me to do something with her. It was impossible to concentrate and I forgot so many things. The atmosphere was so tense, I was crying in front of my daughter. I had to talk about the domestic violence that I had experienced and my daughter heard everything.[28]

During the interviews themselves concerns were raised that 'decision makers are still prone to disbelief without foundation, and to treating the asylum interview and decision-making process as adversarial rather than as an exercise of an international protection obligation'.[29] During screening interviews conducted at Lunar House interviewing officers were reported to have 'continuously looked at [their] computer rather than listening to the interpreter',[30] and to have 'not allow[ed] the asylum seeker to tell his story but kept intruding with questions'.[31] What is more, the fact that 'Interviews constantly make you relive your trauma',[32] was apparently not enough to sensitise interviewers about the way they asked questions. 'During the interview, I was asked 8 times to give the date that my son died', one asylum seeker quoted in a submission recalled,[33] whereas another described how her 'account of torture as a woman was not believed at all'. She continued:

> I looked like I was wasting the time of the officers. I had nothing special to tell them, they looked like they already knew what I am going to tell them, from

the start till the end. I felt very uncomfortable as if they could not believe a human being could be subject to a torture, it was like I was telling them something from a movie.[34]

Compounding these difficulties is a strong sense of distrust of government officials and interpreters among many asylum applicants. 'In our country', one former asylum seeker told me during an interview, 'the interpreters work for the government, so for me I didn't know if the interpreter worked for the government or for charity or what? And I didn't know if what he was saying was true or not.'[35]

Within some communities of asylum seekers waiting to be interviewed, it only takes one account of poor treatment within Lunar House, or the regional offices where substantive interviews have been conducted since the late 2000s, to provoke significant anxiety about forthcoming interview events: one harsh word or angry gesture from an interviewer or member of security staff can reverberate widely around networks of asylum seekers and their supporters. The same Kurdish asylum seeker who had to wait eight years for a final decision described his anxiety upon entering Lunar House and being interviewed because of the accounts of deportation he had heard prior to his arrival:

I was with one of my friends. Cause he drived there, so he was staying in the car and I said if I didn't come out just ring my family, my friends, the human rights activists in our area. [Because] the thing is you think you are not going back, that something is happening inside. Lots of people say before. People said 'Okay if you go into that building, you maybe not coming out, they can take you to the airport and send you back.' So that was scaring everyone.[36]

When applicants are anxious and uncomfortable it is clear that the social atmosphere of the interview is far from ideal. 'Inside is scary', he explained, 'because inside there is all these cameras everywhere. It was like a prison. I feel very scared'. Rumours were also not limited to accounts of deportation (which may have had some truth in them[37]) but extended to far-fetched accounts about how the Home Office assesses applicants. 'In the interview there was four people there', he recalled in describing his substantive interview,

my solicitor and then two people were sitting and one guy standing in the room. And I thought 'Why is he standing?' 'cause I'd heard a story that some people are standing inside the interview room and they are there to watch you, what are you like, are you worried about the process. And that person is a psychological security person. And always in my head I was asking myself 'Is he a security guy or is this a psychologist guy?'[38]

Allport's fifth condition for meaningful, prejudice reducing interaction, concerning equal status, is perhaps the most comprehensively violated of all his conditions in the context of asylum interviews. Jean-Francois Bayart (2007) has argued that one of the defining features of the modern globalised world is the way in which poor and marginalised groups are forced to wait. From rising prison populations in the United States and parts of Europe, to increasing displaced populations in temporary-come-permanent camps around the world, waiting has been fused with subaltern identities that indicate non-belonging and non-citizenship. Bayart argues that this type of waiting is associated with the rise of the modern state whose role is to police and regulate globalisation and the flows of people that it engenders. Waiting has become part and parcel of the regulation and bureaucratisation of the modern global system. '[P]eople have been incited by powerful institutions to believe in particular visions of the future yet lack the means to realize their aspirations', Jeffrey (2010, p. 3) explains. This results not only in dead time, where all a person imagines they are doing is waiting, but also is accompanied by stigmatisation and insecurity as mainstream populations associate those who are forced to wait with loitering, surplus and idleness (Mbembe, 2004).

For asylum seekers, waiting for the opportunity to make a claim, tell their story fully, or for a decision on their asylum claim is a novel form of torment that has been described as 'permanent temporariness' (Bailey et al., 2002) and a painful 'state of limbo' (Hyndman and Giles, 2011, p. 362; see also Conlon, 2011). 'The uncertainty of the future makes decisions impossible', Webber (2012, p. 56) writes, 'how to study, furnish a room, buy clothes, make friends, plan a future when everything is provisional and insecure? ... this limbo is another torture' (Webber, 2012, p. 56). The same Kurdish asylum seeker who had to wait eight years for a decision described this ordeal as 'the biggest challenge of my life':

> Because you are finding yourself getting older. Every morning you are looking to see if the letter or something has arrived which is really hard and if you're not careful and you are not integrated into society you can have a huge mental health problem. We, as Kurds, we say that if we had stayed in Iraq Saddam was gonna kill us fast, that is the killing process, but here they kill you by keeping you waiting. Sometimes people kill themselves. They think that this life is nothing. I always described myself, before getting status, as a baby in the stomach of the mother. Not born. As soon as you are born, as soon as you get status, your life starts. Because when you get status every door opens for you. But before, you cannot plan. You say 'I want to buy a house', no you don't have status, it could be any time you get refusal and they send you back. I wanted to make an investment, you can't. I wanted to start life providing for my family but all these processes are stopping you.[39]

This type of waiting is the result of more than simply bureaucratic delays. It performs the authority of the dominant group and asserts the status of applicants by demonstrating how low down the list of priorities migrants are. It also forces asylum seekers to perform their own subaltern identities: they hope, they long, they speculate, they gossip, they dream, they fixate, they get bored, they talk not of an improvement to their status, but of getting any status at all.

Visits to government immigration offices bring these interminable periods of waiting to a head, and with these visits the requirement to perform migrant submission to the state intensifies. Queues form, 'filled with feelings of anguish, humiliation and attitudes of resignation' (Pérez, 2010, p. 168), which represent, above all, 'the basis of the social organisation of access and delay' (Pérez, 2010, p. 168). At Lunar House in the 2000s the queues sometimes started at 5.00 am, often reached five hours in length and regularly held over a thousand people in draughty, semi-covered areas, many waiting to lodge an asylum claim and undergo the screening interview. 'What I saw when I went there', recalled one activist who had accompanied an asylum seeker to Lunar House,

> were traumatised people, elderly women, pregnant women being herded and being forced to stand around in a way that it's not right for people when they're old and carrying children to be forced to stand in queues, to be treated as though they have no rights at all. I was totally horrified.[40]

She continued:

> we have rules and regulations about how we treat animals. We can't make them stand around and not give them water and punish them, we're not allowed to do that to animals, and actually what people were getting there was just appalling, absolutely appalling. They're victims, there's nothing they can do for themselves, they are totally dependent on people's good offices, they are totally dependent on people doing their job with care and compassion and actually that was absent.

Similar concerns were reflected in the SLC report. 'Please make [the] waiting area humane', one respondent pleaded (Back *et al.*, 2005, p. 16). '[The] queue area is icy cold', wrote another, '[i]t makes people sick' (Back *et al.*, 2005, p. 16).

Applicants were also humiliated while they waited: provision for women with children, for example, was particularly bad. 'If you have a baby you can't even get nappies', one activist told me. 'You can find nowhere to change a baby and it causes huge distress for people standing in queues for hours with crying babies, with no support or facilities at all.'[41] Others reported having to carry pushchairs up to the second or third floor of the building.

Security staff sometimes exacerbate the difficulties. 'There was this pregnant woman next to me', one activist told me, 'who was leaning against something and [the male security guard] came out and he just abused her and told her that she wasn't allowed to lean. It was dreadful.'[42] 'You feel it when people don't like you,' a former asylum seeker explained,

> they have a way of talking to you. When you ask for something, for example. I asked 'Where's the toilet?' The way she answered me was…[shakes head]… and she just said 'wait here' and the security guys come and take you to the toilet. And I said 'you show me where and I will just go there', but they wouldn't.[43]

The very design of Lunar House establishes the (non-)status of those who wait. 'There is something about the physical architecture of that place, the things that are literally set in stone and in glass that reveal so much about the way the government and the state views asylum and immigration',[44] an activist observed. 'It's like a DHSS[45] waiting room but not as comfortable. People are forced, physically, to sit on the edge of their seats. Sometimes talking about the most personal, difficult, traumatic experiences of either persecution or what it meant to flee that persecution. And they are doing it literally sitting on the edge of a metal chair that is bolted to the floor, they have to lean forward.'[46]

The metal chairs were part of the security measures taken within Lunar House. As another activist explained, 'when we put it to one of the senior people of the IND "Why do these chairs have to be bolted to the floor?" he said "well it stops people throwing them through the window".'[47] These sorts of security measures, however, send a message to the applicants that

> we're frightened of them. So much that we have to fix their chairs to the ground, we have to put a protective screen in place. It creates a culture of suspicion which the recipients feel and which makes them feel hostile because they are being treated as though they are not really good people.[48]

In the worst cases, the security measures can themselves be provocative. Applicants recalled how they were required to 'sit to attention' (Back *et al.*, 2005, p. 21) during their screening interviews and how difficult it was to hear the interviewer because of the protective plastic screen that separated them. 'The whole system makes them feel threatened', one activist explained. 'If you're a member of staff and you've had five people have a go at you in one morning, and really get aggressive, you need that screen. But why did those people get aggressive in the first place? Because the screen was there! They couldn't speak properly! The seats are so far away [that] everybody can hear…their personal business.'[49]

Psychologists have argued that emotions are centrally important in reducing prejudice and we will revisit emotions in greater detail in Chapter Six. But it is worth reflecting here on the particular way in which anxiety is spatially managed and produced in Lunar House. Anxiety reduction between two groups 'has been posited as *the* [italics in the original] mechanism through which contact elicits improved inter-group relations' (Stephan and Stephan, 1985; Hewstone, 2003; Pettigrew and Tropp, 2008; Matejskova and Leitner, 2011). Conversely, 'by far the strongest [mediator of positive contact] is intergroup anxiety' (Barlow *et al.*, 2012, p. 2). In other words, the surest way to poison a contact event, according to contact theorists, is for the parties to be suspicious and fearful of each other. Furthermore, where suspicions are confirmed and contact is negative – perhaps because of a hostile or aggressive meeting – these experiences of negative contact *are proportionately more powerful in nurturing prejudice than positive contact is in eradicating it.* 'The beneficial effects of numerous positive intergroup encounters', Barlow *et al.* (2012, p. 12) explain, 'may be counteracted by the relatively infrequent but powerful effects of negative intergroup encounters.' The result is that negative, hostile and aggressive meetings are actually a more powerful predictor of prejudice than positive contact (Paolini *et al.*, 2010). To design the setting in which applicants meet decision makers in anticipation of negative contact, thereby raising the fear, expectation and likelihood of aggression and hostility on both 'sides', is therefore profoundly counterproductive to the nurturing of meaningful, prejudice-reducing interaction.

In summary, it is difficult to imagine how both the screening and substantive interviews that were conducted within Lunar House during the 2000s could have been staged in a way that was less conducive to meaningful, positive contact and morally obligating encounters. The majority of Allport's conditions are violated at the site, which still hosts most screening interviews of applicants applying for asylum from within the United Kingdom. These conditions include the stipulation that the parties work towards similar goals, that the contact is sustained, that the interaction occurs in a relaxed and free-flowing way and that the status of the parties is roughly equivalent. It is the specific organisation of contact spaces and settings that undermines many of these conditions: the case of Lunar House confirms the importance of 'the spatial and temporal contexts within which prejudice occurs' (Valentine, 2010, p. 534). It illustrates not only 'how space and place influence the kinds of interactions that take place between state agents and citizens' (Jones, 2012, p. 819) but also the way in which the 'detail and texture of interaction' (Askins and Pain, 2011, p. 816) is conditioned by institutional arrangements.

Appeals

It is a telling measure of the inadequacy of the interviews and the initial decision making process that around a quarter of negative initial decisions are overturned on appeal, indicating that at least a quarter of them are

Figure 4.4   First-tier Immigration and Asylum Tribunal Hearing Centre, Columbus House, Newport, Wales. Photograph by Melanie Griffiths, 2013.

erroneous[50] (Amnesty International and Still Human Still Here, 2013). This is not to say, however, that the appeal process is any less encounter-averse. It involves hearing roughly 15,000 asylum appeals per year in 13 regional tribunals[51] dotted across the United Kingdom, such as the one in Newport, South Wales (see Figure 4.4), and has been criticised for lacking legal coherence (Thomas, 2011) as well as obscuring a set of 'underlying power relations' beneath a veneer of 'impersonality and neutrality' (Campbell, 2009, p. 9). The real value of asylum appeals, Thomas argues, is their 'symbolic and expressive functions' (Thomas, 2011, p. 23) which provide powerful techniques for appeasing individuals and quietening political opposition, but do not actually offer 'much by way of legal substance' (Thomas, 2011, p. 23).

Here I draw on a 3-month, multi-sited ethnography of hearing centres conducted in 2013, that took in eight of the thirteen major hearing centres in the United Kingdom, to assess whether appeals do a better job of meeting Allport's conditions than the interviews at Lunar House (see Appendix for more details on methodology). The data I draw on in this section are taken from the ethnographic research diaries of my researchers Drs Melanie

Griffiths and Andrew Burridge. Where the first person is used in direct quotations in this section it refers to one of these two researchers, who I denote with the initials MG and AB respectively.

Given the lack of substance of appeals, the principle of fairness is consistently at risk of being overwhelmed by administrative considerations. In particular, the Home Office is very concerned that the length of the appeal process may make it difficult to eventually remove failed applicants, and therefore tribunals and judges themselves are under pressure to complete decisions within 10 days of every hearing, and conclude 75% of asylum appeals within 6 weeks (Thomas, 2011, p. 98). This has the effect of putting pressure on the length of hearings and ensuring that sustained interactions in the form that Allport advocates do not occur. Actors involved in cases are especially keen not to 'try the patience' (MG[52]) of the immigration judges by prolonging the hearings. In one case that was observed

> both [sides] expressed a desire to get the case over and done with as quickly as possible and both were extremely brief in their roles. The whole hearing is just 48 minutes long!! And there are two witnesses asides from the applicant! I tell [the legal representative for the Home Office] afterwards that I'm shocked how quick it was and she replies by telling me that that wasn't particularly quick, she can do two asylum appeal hearings by 11.15 am![53] She's clearly proud of this, feeling that speed is a good sign. (MG[54])

In terms of the pursuit of common goals, the legal approach used during asylum hearings in Britain is adversarial meaning that the two parties will seek to undermine each other's credibility through a legal struggle fought by opposing sides that has traditionally been seen as a 'trial of strength' between two adversaries rather than an enquiry into truth (Rock, 1993, p. 31, citing Devlin, 1979). Asylum seekers invariably enter this battlefield on an unequal footing because institutions such as the Home Office are better equipped to devise new legal strategies (Campbell, 2009). Some applicants may have lawyers whereas others will not, partly as a result of the systematic contraction of legal aid made available to asylum seekers in this situation in recent years.

For its part, the Home Office usually fields a Presenting Officer (PO), whose job involves 'having to be combative and accuse people of lying. Being too "soft" or trusting [is] considered a problem' (MG[55]). These officers are usually relatively junior civil servants 'with no legal qualifications' (Good, 2007, p. 112), who commonly focus on 'trying to induce appellants to make inconsistent remarks' (Good, 2007, p. 112). One PO described,

> how difficult her job is, how demoralising and what pressure Presenting Officers are under. She said that they are undertrained and underpaid (in comparison to trained legal representatives in particular)...They are

expected to win 60% of their cases but that this is unfair because some cases are bound to be successful, no matter how good the Presenting Officer. [I]f they don't meet their targets, they get a talking to (MG[56]).

In contrast to POs who fall short of their target, Officers who do well and either meet or exceed their target 'success' rate (meaning that the judge decides to uphold the initially negative decision of the Home Office and refuse the asylum claim) are rewarded with petty bonuses like high street shopping vouchers and extra holidays (Taylor and Mason, 2014). Conversely, those at risk of losing cases are also encouraged to withdraw them rather than notch up a failure against their record, prolonging the agony of waiting for many applicants with strong cases.

The net effect is that it is hard to conceive of a situation in which the interests of the two parties are more diametrically opposed. So pronounced are the incentives to undermine each other in the legal context that even the most basic forms of body language and linguistic etiquette are impacted, to the extent that the social atmosphere and ease of communicative interaction in the tribunal are often extremely negative and impaired. During hearings POs often purposefully keep their heads down or in their hands and almost never look at the appellant (MG[57]). One Officer recognised that 'calling someone a liar to their face is very difficult and not something you would ever do in normal society' (MG[58]). POs consequently seemed to 'deal with the emotions of their job either by being very aggressive to the appellant, or creating a barrier with them, which results in their not looking at the appellant (e.g. closing their eyes when talking to them, staring at the table or their documents, addressing the translator instead, or even looking in the opposite direction)' (MG[59]).

In terms of spoken interaction, the constant drive to find inconsistencies in applicants' accounts can result in the complete breakdown of effective communication. POs at one centre, for example,

> seem to revel in miscommunication. Unlike in everyday communication, where both speakers attempt to repair the conversation when there is misunderstanding, it seems to be a strategy of POs to allow miscommunication to persevere, whether because it frustrates judges or appears as though the appellant is being evasive or difficult. So, if an appellant doesn't appear to understand the question (but attempts to answer it, rather than say they don't understand), POs will either repeat the question using exactly the same words (rather than try to explain themselves in another way), but with growing anger, or just accuse the appellant of not answering the question. (MG[60])

These stilted forms of bodily and verbal interaction are made worse by a strong culture of separating the parties before and during hearings. Immigration judges are segregated from applicants via separate offices and corridors, reflecting long-standing concerns over 'contamination' by the witnesses and applicants before their moment in court (Rock, 1993). Tribunals are consequently designed in order to segregate different groups involved in hearings, and only a few types of individuals (e.g. security staff and ushers) are able to traverse the different areas. The rift between POs and applicants is most obvious in the common waiting areas of the tribunals:

> The Presenting Officers either wait upstairs in their office, in the reception area talking to Legal Representatives or ushers, or walk briskly through the waiting room to the hearing rooms. Presenting Officers are usually the first ones in the hearing rooms. [They] *never* linger in the waiting room, only ever going through it quickly. Presenting Officers do not like to mix in these public spaces. (MG[61])

Overlaying these impediments to free-flowing interaction is a set of indicators of the unequal status of the actors involved in the hearing. While the design of the tribunal rooms clearly bestows the judge with highest status (they sit on a raised dais, under a coat of arms, and have their own door into the courtroom), the legal representative and the Presenting Officer are usually closest to the judge, with the applicant being furthest away (Figure 4.5). All the actors in the room must address the judge as Sir or Ma'am, stand when the judge enters or leaves, and it is good practice to also bow slightly upon leaving a tribunal room if the judge is still present. Although these rules of etiquette apply equally to all the actors in the case (including the translator if present, the applicant, their legal representative if present and the PO), it

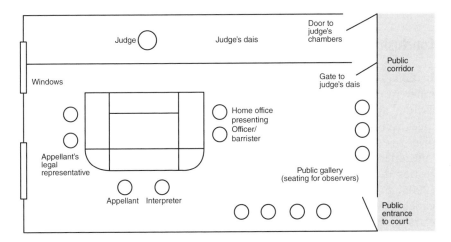

Figure 4.5   Layout of a Tribunal hearing room. Credit: Dr Melanie Griffiths for the original sketch and Dr Rebecca Rotter for the computer-generated version.

is most frequently the applicant who has not been told about the rules, has not understood them or is most unfamiliar with procedures, thereby being the most frequently embarrassed by having to be reminded of them. During verbal communication other rules apply. The applicant should address the judge, for example, even if they are speaking through an interpreter, and speak slowly in short sentences to allow the judge to make notes. The interaction is therefore punctuated by frequent periods of silence as the judge writes down what is happening before signalling for interaction to continue. Often judges will be annoyed by infringements of such rules and although most remain calm and reasonable, sometimes this will result in the applicant feeling the heat of the judge's anger. Some judges who feel that the applicant is not cooperating can become impatient, frustrated and rude, and are more or less at liberty to do so in their own courtroom. This contrasts starkly with the prevailing attitude that, in relation to applicants themselves, 'the court is not a place for emotion' (AB[62]) and those who show emotion are 'generally seen as making the day much harder. ... There is no space for appellants to be angry or frustrated' (MG[63]).

All in all, while the appeal process constitutes an indispensable check against poor quality initial decisions such as those that might be made at Lunar House, it nevertheless repeats the violations of Allport's conditions for meaningful, prejudice-reducing interaction committed during asylum interviews. Although, like interviews, appeal hearings are legally sanctioned and therefore supported by relevant authorities, they are one-off events that lack the frequency of interaction that Allport described. They are also highly formalised exchanges that embody distinct asymmetries of status and, perhaps most importantly, are arenas in which parties strive explicitly towards diametrically opposed goals.

Conclusion: On the Arts of Averting Encounters

In Chapter Two I explored the argument that bureaucracies have an inherent propensity to keep bureaucrats and subjects apart. In the case of asylum application determination, however, functionaries also have to meet asylum applicants at various stages of their application process. This gives rise to a dilemma: how to meet asylum applicants without becoming embroiled in the details of their experiences or affected by the harrowing stories they have to tell. In other words, how to nurture what Barnett (2005, p. 10) has called 'strange proximity'. Functionaries have to be capable of meeting asylum seekers without compromising the moral estrangement that allows them to act indifferently, objectively and dispassionately according to the rules of immigration control rather than their own codes, mores and sensibilities.

This ability is secured by the institutional settings in which these events take place, which facilitate adiaphoric, morally disinterested meetings between officials and migrants by opening various distances between decision makers and subjects aside from physical distance. In the cases of screening interviews, substantive interviews and asylum appeals, contact is infrequent and unsustained, the interaction is formal and stilted, the parties work towards disparate objectives, and the asymmetrical status of applicants and officials overshadows their interaction. This stymied form of interaction, occurring across a range of temporal, social, cultural and organisational forms of distance, leaves little room for meaningful interaction and morally demanding encounters to occur.

System managers may see no need to pursue contact of any other form in the context of border control. They may be so locked-in to bureaucratic, operational ways of thinking that they would question the utility and feasibility of pursuing Allport's conditions at all. Unfortunately, however, the cold, clinical and impersonal systems that this operational mindset produces turn out to be the ideal breeding ground for the sort of individual prejudice and systemically prejudicial decision making that has tarnished the reputations of the IND, UKBA and their descendants over the past 20 years, as discussed in the introductory chapter.

These observations constitute a second critical development of the account of moral distance presented in Chapter Two. Although moral distancing is an important consequence of recent changes in the organisation of border controls, it is apparent from the discussion in this chapter that where physical distance is closed, moral estrangement need not be. There are various barriers to the 'morality inside me' (Kant, 1788/2002) that physical proximity promises to activate. The lack of more encounters like the one between Mr Brimelow and Nataliya, described in Chapter Two, owes itself largely to the existence of these institutional forms of distancing.

The fact remains, however, that both asylum interviews and appeals are, by their nature, occasional and intermittent affairs. Conceivably, therefore, one could defend the British immigration control system by arguing that there is a limit to the degree of meaningful contact that *can* be generated in these situations. According to this line of argument the immigration control system might be merely a victim of circumstance, unable to nurture meaningful or moral encounters between its functionaries and migrants, even if there was a desire to do so, due to the short time they spend together. In the next chapter, however, I examine the organisation of interaction between functionaries and migrants under the conditions of sustained contact that immigration detention produces. As is turns out, the ingenuity through which meaningful interaction and morally demanding encounters are avoided is, if anything, even more striking in this context.

Notes

1 The conversations I had at the counter-protest that I recount in this section occurred on 27 July 2013.

2 The evidence from officials at the demonstration that I quote here comes from my research diary, which I wrote on the basis of scratch notes made during and immediately after the demonstration.

3 There are also Asylum Case Management Review hearings at which parties agree on the issues under consideration in the substantive hearing, and there are also bail, deportation and immigration hearings. I do not discuss these here.

4 The importance of frequent contact for prejudice-reduction, which occurs in a variety of social contexts, has been an enduring finding of the contact literature since Allport's seminal work (Rothbart and John, 1985).

5 Interview with activist, 10 October 2006.

6 Interview with former asylum seeker, November 2013.

7 Interview with activist, October 2006.

8 Interview with senior manager of the IND, Apollo House, 2006.

9 Interview conducted 20 June 2006.

10 Former asylum seeker and activist, April 2006.

11 Interview with activist, October 2006.

12 Afterwards, two A-levels and five GCSEs were required. The A-Level is an abbreviation for the General Certificate of Education Advanced Level, which is an academic qualification offered by educational bodies in the United Kingdom to students completing secondary or pre-university education. The General Certificate of Secondary Education (GCSE) is an academic qualification awarded in a specified subject, generally taken in a number of subjects by students aged 14–16 in secondary education in England, Wales and Northern Ireland. Neither of these represents a degree level qualification.

13 Interview with former intelligence officer, 28 November 2013.

14 Figley (1995, p. 7) defines secondary trauma as 'the stress resulting from helping or wanting to help a traumatized or suffering person'. See also secondarytrauma.org for a list of the common symptoms of secondary trauma as well as a list of professionals who are at risk from it.

15 Interview with activist, 20 June 2006.

16 The quote is taken from a 'PCS View' web article published in 2003 and quoting the branch secretary with responsibility for Lunar House (Public and Commercial Services Union, 2003).

17 Interview with former intelligence officer, 28 November 2013.

18 The officer herself identified as British Minority Ethnic (BME).

19 Interview with former intelligence officer, 28 November 2013.

20 At the time of my primary research at Lunar House during the mid-2000s substantive asylum interviews were usually held in Lunar House or the surrounding area of Croydon. With the creation of Regional Asylum Teams (RATs), however, substantive interviews began to be undertaken at other locations including Folkestone, Cardiff, Solihull, Liverpool, Leeds, Newcastle, Glasgow, Belfast and elsewhere in London.

21 Interview with senior manager of the IND, Apollo House, 2006.

22 Interview with activist, October 2006.

23 Interview with senior manager of the IND, Apollo House, 2006.

24 Interview with former asylum seeker, November 2013.

25 The Home Affairs Select Committee is a committee of the House of Commons in the Parliament of the United Kingdom, with responsibility for examining the expenditure, administration and policy of the Home Office and its associated public bodies. The Committee published its report entitled 'Asylum' (Home Affairs Select Committee, 2013a) in October 2013 after a 10-month inquiry into the asylum system. The report highlighted the slow pace of decision making, the poor quality of initial decisions such as those made at Lunar House, the poor quality of housing for asylum seekers provided under new contracts with private contractors, and the inadequacy of financial support for asylum seekers and those who have been refused but are unable to return home. As part of researching the report, the Committee invited submissions of evidence and many asylum support groups responded.

26 This evidence is taken from the written submission by Leicester City of Sanctuary to the Home Affairs Select Committee (2013b) on Asylum.

27 This evidence is taken from the written submission by Wyon Stansfeld to the Home Affairs Select Committee (2013b) on Asylum.

28 This evidence is taken from the written submission by Women's Refugee Strategy Group to the Home Affairs Select Committee (2013b) on Asylum, and the woman cited is from Pakistan.

29 This evidence is taken from the written submission by ASSIST Sheffield to the Home Affairs Select Committee (2013b) on Asylum.

30 This evidence is taken from the written submission by Lucy Fairley to the Home Affairs Select Committee (2013b) on Asylum.

31 This evidence is taken from the written submission by Lucy Fairley to the Home Affairs Select Committee (2013b) on Asylum.

32 This evidence is taken from the written submission by Lucy Fairley to the Home Affairs Select Committee (2013b) on Asylum.

33 This evidence is taken from the written submission by Women's Refugee Strategy Group to the Home Affairs Select Committee (2013b) on Asylum, and the woman cited is from the Democratic Republic of the Congo.

34 This evidence is taken from the written submission by Survivors Speak Out to the Home Affairs Select Committee (2013b) on Asylum.

35 Interview with former asylum seeker, November 2013.

36 Interview with former asylum seeker, November 2013.

37 Lunar House has its own short-term holding facility in which asylum seekers are held, usually for a few hours, before being taken into detention from where they can be deported.

38 Interview with former asylum seeker, November 2013.

39 Interview with former asylum seeker, November 2013.

40 Interview with activist, 28 February 2006.

41 Interview with activist, 28 February 2006.

42 Interview with activist, 28 February 2006.

43 Interview with former asylum seeker, November 2013.

44 Interview with activist, October 2006.

45 The Department of Health and Social Security (commonly known as the DHSS) was a ministry of the British government from 1968 until 1988. It

gained notoriety for its drab and uncomfortable waiting rooms, as depicted in Paul Graham's photograph entitled *Crouched Man, DHSS Waiting Room, Bristol,* which can be viewed here: http://www.moma.org/collection/object.php?object_id=55222 (accessed 31 August 2015).

46 For a discussion of this phenomenon, as well as other aspects of both Lunar House and the SLC report, see Back (2007).

47 Interview with activist, October 2006.

48 Interview with activist and community leader, October 2006.

49 Interview with activist and community leader, October 2006.

50 It should be noted that at this point in the argument it is necessary to take up some of the language and concepts of the Home Office, which ordinarily I would object to. The notion that asylum claims can be 'erroneous' only makes sense within the terms set out by the Home Office, and a more radical position would question whether anyone should be deemed erroneous on the basis of their attempts to travel for a better life.

51 Tribunals operate like courts but have a slightly less formal status in law.

52 Diary entry of 28 August 2013.

53 Immigration tribunal hearing schedules in the first tier generally begin at 10.00 am.

54 Diary entry 17 October 2013.

55 Diary entry 29 October 2013.

56 Diary entry 28 August 2013.

57 Diary entry 23 August 2013.

58 Diary entry 23 August 2013.

59 Diary entry 29 October 2013.

60 Diary entry 29 October 2013.

61 Diary entry 9 September 2013.

62 Diary entry 8 November 2013.

63 Diary entry 29 October 2013.

Chapter Five
Indifference Towards Suffering Others During Sustained Contact

I have never been able to understand how it was possible to love one's neighbors. And I mean precisely one's neighbors, because I can conceive of the possibility of loving those who are far away. I read somewhere about a saint, John the Merciful, who, when a hungry frozen beggar came to him and asked him to warm him, lay down with him, put his arms around him, and breathed into the man's reeking mouth that was festering with the sores of some horrible disease. I am convinced that he did so in a state of frenzy, that it was a false gesture, that this act of love was dictated by some self-imposed penance. If I must love my fellow man, he had better hide himself, for no sooner do I see his face than there's an end of my love for him.

> Spoken by the character Ivan Karamazov in Fyodor Dostoevsky's
> *The Brothers Karamazov*, and cited in Wendy Hamblet, 2003, p. 361

[D]irect experience establishes too close a contact
> Hannah Arendt, *Essays in Understanding*, 1994, p. 323

I began this book with an account of Mr Dvorzac's death in immigration detention in 2014, which was notable because it was not distant or remote from functionaries but occurred right under their noses. Although moral distance can help to explain some forms of indifference towards suffering, it struggles to shed light on indifference that persists in these sorts of circumstances. In the last two chapters I have critically reflected on some of

Nothing Personal?: Geographies of Governing and Activism in the British Asylum System,
First Edition. Nick Gill.
© 2016 John Wiley & Sons, Ltd. Published 2016 by John Wiley & Sons, Ltd.

the implications of the account of moral distance I put forward in Chapter Two, arguing that moral distance needs to be augmented by an attention to the distancing of both subjects and bureaucrats (in Chapter Three) and with an attention to the ways estrangement can be maintained at close quarters (in Chapter Four). In this chapter I venture further beyond explanations couched in terms of moral distance to seek to account for Mr Dvorzac's death. The sort of indifference generated through moral distance is unfamiliar with suffering, but here I argue that overfamiliarity with suffering can give rise to a different form of indifference.

Georg Simmel (1903/2002) first linked the development of a blasé attitude towards others with modern urban life. Contemporary cities, he argued, throw people together with such frequency, and bring diversity into such proximity, that the modern, metropolitan personality 'creates a protective organ for itself against the profound disruption with which the fluctuations and discontinuities of the external milieu threaten it' (Simmel, 1903/2002, p. 12). Modern individuals consequently develop a way of relating to others that is characterized by 'a purely matter-of-fact' (Simmel, 1903/2002, p. 12) attitude, 'unrelenting hardness' (Simmel, 1903/2002, p. 12) and a 'blasé outlook' (Simmel, 1903/2002, p. 14) that tends to result in 'indifference towards the distinctions between [others]' (Simmel, 1903/2002, p. 14). This suggests that our moral impulses, most at home in close proximity, are in danger of being overstimulated and exhausted as modern cities expose us with unprecedented frequency to different people and moral claims. This form of indifference does not rely upon physical distance: it is impervious to nearness, and may even be exacerbated by it.

Clearly in the case of the abused men in Harmondsworth, the sort of moral awakening that Levinas describes in proximity did not occur: the ethical moment of epiphany was somehow suspended despite the sustained contact between guards and detainees. As Dostoevsky intimates through his character, Ivan Karamazov, loving people and empathising with them is hard work, especially when they appear disgusting, inconsequential or strange. 'Empathy involves imagining ourselves in someone else's skin and allowing ourselves to feel some of their suffering by association. This is painful' explains Stansfeld (2013). '[I]f you really open yourself with presence to other people's feelings and needs' Krznaric writes,

> then you might become overwhelmed by the experience, resulting in emotional distress and inaction. ...you can think of [it] as leaping *too far* into someone else's imagination. [Such distress and inaction] has been observed especially amongst those working in emotionally extreme and traumatic situations.
>
> Krznaric, 2014, p. 116

To avoid this risk of becoming overwhelmed, one response when confronted with suffering is to enter a 'state of denial' according to which a variety of psychological ruses are employed to hide the truth about trauma and atrocity (and our implication in them) from ourselves, including feigned blindness to obvious events and extreme fidelity to the technical demands of one's role (Cohen, 2001).

For those that refuse these ruses and instead attempt to remain sensitively and empathetically engaged with trauma victims, the consequences are often dire. It is well-known that those who work closely with traumatised others and acknowledge their suffering are at heightened risk of what has been called compassion fatigue, burn out, secondary traumatic stress syndrome and vicarious traumatisation (Figley, 2002). This phenomenon, which I touched upon briefly in the previous chapter, can result in the emotional contagion of one set of feelings by another individual and, when not properly supported by counselling, '[t]he picture that emerges is clear: Those who work with the suffering suffer themselves because of the work' (Figley, 2002, p. 5). Even professionally trained counsellors and psychotherapists report their struggles with feelings of guilt, anger, stress, frustration, exhaustion, powerlessness and helplessness when working with trauma survivors (Century et al., 2007; Jensen et al., 2013). The options for such professionals are either to face up to the secondary trauma head-on, which requires courage and is emotionally challenging, or to adopt shortcuts to avoiding compassion fatigue including 'detachment [as a] distraction away from the humanity of the patient' (Figley, 2002, p. 218). How much more likely is this detachment among Detention Custody Officers (DCOs) who have not been trained or prepared anything like as well as these professionals?[1]

One way to interpret the forms of indifference responsible for the poor treatment of detainees is to repudiate theories of the encounter. Wendy Hamblet employs a literal reading of Levinas and takes issue with his conceptualisation of ethics and the status that he accords to the face to face encounter on these grounds. 'Emmanuel Levinas's phenomenology, offering a naïve description of the way in which humans occupy their subjective "places" on earth, composes a remarkably – and disturbingly – sympathetic account of living being', she argues (Hamblet, 2003, p. 356). Far from responding with generosity in the face of suffering,

> [r]evulsion in the face of the nearby needy resonates with most of us, though we may be ashamed to admit it. Needy strangers on the streets of our cities are most often met with fear or disgust. The unknown other is terrifying, horrifying; his difference is experienced as threatening, precisely in proportion as his abjection is beyond question.
>
> Hamblet, 2003, p. 361

'Nearness frustrates compassion' (Hamblet, 2003, p. 363), she surmises, in opposition to Levinas. 'The ugly reality of the miseries of the abject – their foul smell, their louse-ridden and torn rags, their wary looks and suspicious demeanour – render impossible the "identification" from which ethical response might be launched' (Hamblet, 2003, p. 363). The horrific essence of immigration detention, on this logic, lies in recognising this impossibility. By exposing functionaries to innumerable disturbing stories of past individual traumas as well as legion examples of present suffering they quickly become overwhelmed, their sympathies exhausted, and they are obliged to spontaneously develop psychological techniques of estrangement, detachment and aloofness – i.e. indifference – as a form of self-care.

Whether or not we follow Hamblet's rather literal interpretation of Levinas, her broader points about disgust and revulsion, and consequent psychological detachment, are hard to ignore and are in keeping with Simmel's discussion of the blasé urban personality. Disturbingly, it is possible to discern particular institutional tactics that make this sort of detachment among detention personnel both imperative on the one hand and easier on the other. As I show in the following sections, the spatial churning of detainees from centre to centre exposes staff to a greater diversity of suffering and forces them to step back from close empathetic relationships with individual detainees, while the institutionalised trivialisation of detainees, combined with their often-asserted strangeness, makes it easier for staff to convince themselves that refusing to respond sensitively to detainees is acceptable.

The most striking aspect of these institutional features is the way in which contact turns into something that actually feeds and exacerbates the indifference that staff exhibit. In the case of the sort of psychological avoidance that Simmel describes, the more exposure to trauma the more pronounced and developed the avoidance tactics are likely to be (Figley, 2002). In a reversal of the relationship between indifference and contact to that described in previous chapters, therefore, under the extreme situation of sustained contact in immigration detention, contact with suffering others itself acts to jam, clog and overwhelm the empathetic psychological capacities of functionaries. This is why it is imperative to appreciate that indifference is of at least two highly distinct forms: that derived from distance and unfamiliarity, and that derived from over-closeness and overfamiliarity. It transpires that sustained contact with suffering others often has an opposite relationship to the two: promising to alleviate the first but threatening to exacerbate the second.

In what follows I explore the limits to the ethical potential of the face to face encounter that are set by overfamiliarity with suffering and overcloseness to the 'ugly reality of the miseries of the abject' (Hamblet, 2003, p. 363). I first set out the background to immigration detention in the United Kingdom, characterise the role that Detention Custody Officers play in Britain's removal centres, and establish the riskiness of immigration detention to the British state by pointing to a set of situations in which empathetic, morally proximate relationships *have* developed between staff and detainees (see also Hall, 2012). I then explore the institutional arrangements that either make the need to develop a blasé, insensitive attitude towards detainees more pressing, or make it easier to do so, and conclude by assessing the cruel consequences of the insensitivity that is thus engendered.

Immigration Detention in the United Kingdom

At the time of writing the United Kingdom utilises 11 detention centres with a capacity for around 4000 detainees at any one time. Additionally, so-called 'pre-departure accommodation' is available at a specialist centre called 'Cedars' in London, which holds families including children (to which we shall return in the next chapter) and there is capacity to hold more individuals at ports and in mainstream prisons. In total, over 30,000 migrants are held under immigration powers in the United Kingdom annually, but despite the scale of the operation immigration detention is largely hidden from public view and often only comes to the public's attention through reports of crises in detention centres in the media, such as the 2002 Yarl's Wood fire, the riots in Harmondsworth in 2004 and 2006, hunger strikes and protests by detainees, and the occasional death of a detainee that is particularly disturbing. The size of the immigration detention population in the United Kingdom has grown steadily from a capacity of 250 in 1993 (Bacon, 2005). All removal centres are contracted out, either to Her Majesty's Prison Service or one of a small number of private security companies such as Serco, G4S, Mitie and Geo. In 2006 I spent time researching the oldest detention centre in the United Kingdom, Campsfield House Removal Centre, and was able to gain access to the centre and the facilities and interview a variety of staff as well as former detainees and activists (see the Appendix for more details).

Detention centres in the United Kingdom are officially run under the Detention Centre Rules (Immigration and Nationality Directorate, 2001) although successive inspections by Her Majesty's Chief Inspectorate of Prisons (HMIP) of detention conditions have revealed that some of the most fundamental principles of these rules are not adhered to in practice. The rules state that:

> The purpose of detention shall be to provide for the secure but humane accommodation of detained persons in a relaxed regime with as much freedom of movement and association as possible, consistent with maintaining a safe and secure environment, and to encourage and assist detained persons to make the most productive use of their time, whilst respecting in particular their dignity and right to individual expression. [...] Due recognition will be

given at detention centres to the need for awareness of the particular anxieties to which detained persons may be subject and the sensitivity that this will require, especially when handling issues of cultural diversity.

Immigration and Nationality Directorate, 2001, p. 4, sections 3(1)–3(2)

In contrast to the tone of these principles, immigration detention in the United Kingdom has been alleged to be the site of secondary torture for torture victims (Medical Justice, 2012), where foreigners who offer no political mileage to politicians can be locked away out of sight and out of mind for an indefinite period under executive powers, that is, without trial or a specified release date (Bail for Immigration Detainees, 2009). 'I think detention system is very hard' one of my interviewees explained,[2] who had spent around 18 months in detention at various centres in the United Kingdom, 'because the people they left, they left our country because of problem and when they come here it's like same, same treatment. So I mean like me I been in my country I been in prison and here I been in detention. I think almost it's the same.'

The United Kingdom is one of only a small number of countries to practise indefinite detention. In the late 2000s France and Cyprus shared a 32-day limit over the length of time that a person could be detained under immigration rules, Italy and Spain 40 days, Ireland 56 days, Portugal 60 days and Germany 18 months (Phelps, 2009). Not knowing the term of their confinement is without doubt one of hardest and cruellest elements of the British detention system. 'The lack of information impacts on their general wellbeing', one activist who regularly visited the detainees in Campsfield told me.[3] 'They start getting frustrated and angry and depressed all at once because they don't know how long they're here for. And a lot of them don't know why they're here either because they don't understand the letters they're given.' It is estimated that nearly 11% of individuals entering detention spend longer than 3 months in detention, and 2% spend more than a year (Home Office, 2011a, cited in Marsh et al., 2012).

Such long spells in detention are expensive. In 2007, despite not normally publishing the costs of immigration detention, the Home Office responded to a freedom of information request to reveal that the weekly cost of incarcerating a single detainee ranged from £511 to £1620 depending upon the centre (The Migration Observatory at the University of Oxford, 2015a). Subsequently, in 2010, in response to a question raised in the House of Lords, it was stated that the average overall cost of one bed per day in the immigration detention estate was £120 (Hansard, 4 February 2010), although since that time it has fallen to around £100 per day. This means that to hold one detainee in detention for a year costs British taxpayers well over £35,000. There are also compensation costs for unlawful detention and other associated legal compensation payments to consider, which totalled £12 million in 2009–10. To make matters worse, the charity Detention

Action has observed that '62% of migrants leaving detention after more than a year are released and only 38% removed' (Detention Action, 2013, p. 1) indicating that for the majority of long-term detainees their incarceration is a futile and wasteful exercise.[4]

The initial decision to hold people in immigration detention is normally taken by an individual, non-judicial immigration officer without formal legal qualifications and without automatic independent oversight (i.e. without an independent actor such as a judge checking the decision[5]). This has given rise to concerns about the arbitrariness of the use of immigration detention (Weber, 2000; Fordham *et al.*, 2013). In particular, one of the central justifications for immigration detention is the perceived risk that the detainee might otherwise abscond, meaning that they could go 'underground' in an attempt to avoid the authorities. Of the small number of studies carried out into the subject, however, Bruegel and Natamba (2002) used the records of Bail for Immigration Detainees (BID), a registered charity, to trace 98 asylum detainees who were bailed between July 2000 and October 2001, finding that over 90% kept their bail conditions despite only 7% receiving a favourable legal decision. This suggests that over 90% of the detainees were incarcerated unnecessarily.

There is evidence to suggest that the standards of care within immigration detention are frequently highly unsatisfactory. In 2013 alone there were 325 recorded incidents of self-harm requiring medical treatment, and there are regular suicides within the detention estate.[6] The concern is that these levels of self-harm and suicide are driven as much by the abusive treatment detainees receive in detention as by their pre-migration experiences. 'They will use horrible terminology and very racist language', one of my interviewees asserted[7] about the guards in Campsfield IRC in the mid-2000s, corroborating the findings of an undercover BBC investigation into detention practices in 2005. Two BBC reporters worked undercover in a detention centre and for a company that escorts asylum seekers and immigrants around the country, revealing evidence of a culture of violence, abuse and assaults against detainees among staff (BBC, 2005c).

By 2012, and despite an independent inquiry following the BBC documentary, the standard of care was still not adequate. Research conducted with 46 detained women in Yarl's Wood detention centre reported that despite the fact that 72% said they had been raped before their arrival in the United Kingdom, 87% had been guarded and watched by a male officer, 70% said this made them feel uncomfortable and 50% said that a member of staff had verbally abused them (Girma *et al.*, 2014). Compounding these issues, health facilities have also been shown to be unacceptably below standard both in general and especially for detainees who are pregnant, who have HIV and who are children (Medical Justice, 2010, 2011, 2012, 2013). 'The health facilities were very poor', one former detainee recalled; 'paracetamol was the most common medication for any disease'.[8]

When detainees complain about conditions to staff (many do not, one staff member noted, because 'they think it will affect their case'[9]), there is concern that these complaints are not taken seriously. 'A lot of the detainees say that it doesn't matter', one activist explained; 'they can make a complaint but nothing necessarily happens once they have made it'. One member of staff who had worked as part of the spiritual support team at a detention centre for over five years alleged:

> If a detainee puts in a complaint about an officer one can be almost sure that the officer's friend will put in a [notification that the detainee poses a high security risk] on the detainee. ... This usually means the segregation block... Senior management have the attitude that detainees are guilty unless proven innocent. Complaints are seldom completed and many more are just swept under the carpet or the detainee moved...we have seen detainees victimised for making complaints.[10]

Most of the detention centres have segregation facilities that provide a useful holding bay prior to deportation or transfer to a different centre. 'Any who react in a "loud" way or any who [the contracting company] think will cause problems, are put into the segregation unit', a spiritual support team member told me.[11] Detainees also report the punitive uses of these facilities. 'Of course they punish you', one asylum seeker who had spent over two years in detention told me,

> they put you some place you have to stay alone. For two days or three days you stayed there. There was some fight with staff. There was a big guy and then they take him and they put him in that place. And it's the same when you have removal if you have removal for like today at 10.00 so at 6.00 they take you to [the segregation unit], so you just waiting for the time to come.[12]

The use of segregation for supposedly high-risk individuals is part of the broader isolation of detainees that detention achieves (see Mountz, 2013, for a general discussion). A survey in 2011 demonstrated, for example, that 49% of detainees in the British detention estate had no legal representation at the time of their interview for bail, and 19% reported never having had an advisor whilst in detention (Bail for Immigration Detainees and the Information Centre about Asylum and Refugees at the Runnymede Trust, 2011). Others have noted their isolation from police protection within detention: 'A detainee has no right to call the police', a former detainee complained, 'even if there are civil disturbance or riots in the centre'.[13] And there are other, less formal, ways in which detainees become separated from legal advice and support networks. 'It is most common to collect people at the weekend when solicitors are off', one support staff member explained[14] with reference to the dawn raids into asylum seekers' private homes that mark the beginning of a period in detention (see Vickers, 2012). 'Another ploy', she continues, 'is to remove their mobile phones so they do not have access to phone numbers. They can, eventually, be taken to check their phone for numbers, but the process causes delay. … Even money has been called "illegal money" and taken from them…'

Unsurprisingly, the social atmosphere within detention centres can be oppressive. Aside from the 'general feeling of uncertainty that [detainees have] no idea what is going on',[15] as one activist put it, detainees often suffer from acute under-stimulation. 'The main thing that people are telling me is just how incredibly bored they are about being locked up in a place and not being allowed to do anything', an activist explained;[16] 'A lot of the people have really high qualifications and they're pretty well educated a lot of them [and they are] just dying from under-stimulation and not having anything to do.' This underoccupation of the detainees combines with a strict etiquette protocol to produce a stifling environment. 'Staff expect politeness', one staff member told me:[17]

> Certain items may be borrowed like hair clippers, TV remote, DVDs, table tennis bats and ball and the detainee's card is handed in as surety. Smoking is only permitted in certain places. Detainees may not have matches, but must ask an officer for a light. Looking too long at the fence is considered a risk and the detainee is likely to be called an escape risk and moved into the segregation cells.

Induction into these regimented surroundings is particularly important in establishing and maintaining order. 'I remember that I was asked to do a translation for an induction for a new guy', one former detainee recalled,[18] 'and he asked me specifically to translate to this new detainee. This guy said "If you give us shit we will give you back more shit." So he asked me specifically to translate this. I think they have a policy of people being subjected to a high dose of fear and not to do any foolish things.'

Empathy in Immigration Detention

Campsfield House IRC is imbued with the hallmarks of the continuing struggle to subdue the detainee population. Although there is free association during the day, meaning that detainees can leave their wings and have access to most of the centre, security is high, including pervasive CCTV monitoring, thorough checking and searching of all visitors to the centre, guards patrolling the 12-foot high, razor-wire-topped perimeter fence, multiple steel security doors and the sanctioned use of forceful control and restraint procedures wherever aggressive resistance is met.

In characterising the relationship between guards and detainees in British immigration detention, Alexandra Hall has noted a tension between 'ideals of impersonality, rationality and formalistic bureaucratic action' (Hall,

2012, p. 144) and the fact that 'the discretionary power that officers [hold] – about designating a threat to security, about the use of force, about protecting the regime – always contain[s] the potential for a response to the detainees that recognises him as more than "detainee"' (Hall, 2012, p. 19). Her ethnographic research within Locksdon[19] Removal Centre revealed evidence of both insensitivity and empathy towards detainees. On the one hand officers were quick to condemn detainees for what they saw as 'over-emotionality' and 'inappropriate' wilfulness (Hall, 2012, p. 169), which they conflated with a 'nebulous and imprecise illegality and moral failing' (Hall, 2012, p. 169). But on the other hand Hall diagnoses occasional 'ethical moments in centre life' at which point 'the boundaries that shaped and emerged from life in detention were dissolved and transcended by expressions of concern and care' (Hall, 2012, p. 20).

Campsfield embodies a similar tension. A variety of structural elements mitigate against compassionate relationships with detainees, including the high turnover of staff owing to the long hours and shift work. '[Guards work] 12 hour shifts, seven days on and a few days off, it's very bad conditions and very bad pay', one activist told me in the mid-2000s;[20] 'They get paid a fraction of what prison guards get paid and there's minimal, minimal training.' 'In general it's clear that you don't need any specific qualifications to work there', another regular visitor to the centre asserted at around the same time.[21] 'There's a lot of ex-prison people and a lot of just general people who just need a job, which I think is quite worrying because if you're going to have detention centres at least they have to be staffed by people who are qualified in dealing with people who've gone through trauma.'

Despite the parameters of the work, though, there is evidence that some of the staff at the centre are willing to help detainees out within the terms of their employment. 'The majority of the officers are good and kind people', one staff member told me,[22] 'just a few are unpleasant'. As one former Campsfield detainee recalled, 'You ask a question of a guard [and they] give you the answer. The guard, he give you what you ask for. Guard help you, this is why Campsfield is different from [other detention centres]'.[23]

Aside from the DCOs themselves other staff, including those on the welfare, spiritual support and education teams sometimes went out of their way to help the detainees. One former detainee described his relationship with a spiritual support worker at Campsfield: 'The last time I go to court', the detainee recalled, 'he wrote a letter to bring to the judge and he supported me, he spoke in the letter, [he] is a good man'.[24] Another staff member admitted that '[t]here is the occasional case where I will stick my neck out … I spoke to [one detainee's] lawyer and she had the case reviewed'.[25] One spiritual support team member listed a catalogue of actions they had taken to support detainees:[26]

We have been able to arrange for luggage to be collected, arranged new lawyers, medical appointments, phone cards, money for some being deported, clothes. We have provided for festivities and have a supply of Bibles, Qur'ans and other faith literature to give away. We have contacted detainees' families, and have put those being sent to NASS accommodation addresses of churches or refugee support groups where they can get help and advice. We have contacted organisations such as Jesuit Refugee Service about those being deported so that they have some support on their return.

Some detainees, however, were suspicious of the help that DCOs and other staff offered. One former detainee who had endured a gruelling four years in detention only to be eventually released into the United Kingdom and awarded indefinite leave to remain (and years later had still not been compensated) was torn over the issue of kindness in detention. On the one hand he recognised that there were some guards in detention willing to act humanely and supportively. 'Some of [the guards] are bad and some of them, if they like you, they'll be nice to you', he reflected,[27] 'That's human nature, if you like someone. They can outwardly [fulfil] their job description but underneath they help me. And there are people like that. Maybe in every centre two or three out of a hundred.' But on the other hand he remained cynical about a lot of the 'help' that detainees received, outlining how guards would sometimes develop favourites and help them as a way to achieve other instrumental goals. In his experience favour would invariably be shown to 'the ones who is helping them out, especially the one who is helping them to deprive other detainees. And they are not short of them as well. They will get some kind of improved wages and better facilities, better room, better things like that.'

Other staff described how they struggled to balance between the desire to help detainees on the one hand and the need to remain objective and dispassionate in their professional roles on the other. 'They are forever needing you to explain a letter to them or just to help them with some pronunciation or help them write a letter', one employee who worked as part of the welfare team explained,[28] 'but I'm wary of getting into that and we're not allowed to offer legal advice.' Indeed, officers were under similarly strict rules. 'They're governed by the same rules as we are', the employee continued,

> so they can be chummy with detainees but if someone says 'I'm going to need a notebook' you can't go and buy him one and bring it in. You can't bring anything in for anyone individually and partly that's for fear of what they call 'conditioning'. So if someone asks you to bring one thing and so next they'll ask you for something big and then they'll ask you for something bigger and then you start feeling that you can't say no. So yeah they can of course be friendly with detainees here and there, and some who've been here for months and months they're sort of old friends and they greet each other with 'All right mate how you doing?' but you can't maintain relationships with detainees.

What transpires is a complex and contingent set of factors that determine whether and to what extent staff display 'kindness' and support detainees.

Although there are a variety of constraints including the rules of the centre, and although the spectre of manipulation and ulterior motives is never far from view, there is also evidence of occasional empathy and compassion.

It is important to note that empathy in itself may not be concerning for centre managers, and indeed the ability to be empathetic features in the job advertisements for detention custody officer roles. The tendency of DCOs to present themselves as 'helpful and caring professionals' has been noted in Mary Bosworth's 20-month ethnographic work in British immigration detention centres (Bosworth, 2014, p. 148). But when empathy causes staff to champion detainees' cases or cast the dehumanising elements of the centre in a bad light it is not to be encouraged by management. In what follows I describe the features of immigration detention that reduce the likelihood that spontaneous human kindness erupts in detention. They act to tip the balance towards insensitivity over 'uncalculated and caring encounters' (Hall, 2012, p. 47) by encouraging staff to, at least, contain their supportive activities within the tight remits of professionalism and dispassion. They can be divided into features that make insensitivity more imperative (and sensitivity more costly) on the one hand, and features that render insensitivity easier to internally justify on the other. Without them we might expect far more kindness in immigration detention.

Making Insensitivity Imperative

Aside from the high staff turnover, detainees at Campsfield and other centres are frequently transferred from one centre to another. As Bosworth (2014) notes:

> Staff in IRCs (via the centre manager) can ask DEPMU [Detainee Escorting and Population Management Unit] to transfer detainees to prison or to another establishment, on the basis of good order and discipline or in response to a request from the detainee. ... Decisions about transfer and admission may also reflect factors unrelated to the person's case
>
> Bosworth, 2014, p. 13

Transfers may be requested by detainees in order to be nearer to family, or to be in the jurisdiction of a first-tier immigration and asylum tribunal that is perceived to be more attractive. But where transfers are unsolicited they can isolate detainees. They also act to curtail the duration of co-presence that staff and detainees share and dilute the depth and quality of relationships. Although the frequency of inter-detention estate transfers is not routinely published, parliamentary questions in the mid-2000s revealed that in 2004/5 the British government spent over £6.5 million simply moving detainees from one secure facility to another within the United

Kingdom (author's calculations from Hansard, 5 December 2005, and Hansard, 9 January 2006). Within Campsfield, '[t]he average length of stay is just 7 days', one education support team worker estimated,[29] although 'if you factor out the overnight stayers it's probably 4 to 6 weeks'. Activists and regular visitors to Campsfield detected an increased frequency of transfers between centres in 2006. 'They've initiated a totally different regime', one activist told me,[30]

> which is that they don't want to leave any detainees anywhere for any length of time. There are a few detainees in Campsfield that have been in there for a long time but very few. They will come in and they will stay for 10 days and they will be moved to Harmondsworth or Yarl's Wood or anywhere but it's a moving process.

Consequently former detainees' accounts of their detention experiences often include tortuous and protracted routes around the detention estate (see Hiemstra, 2013, and Martin, 2013, for similar observations in the US case). 'Before Campsfield House I stay long time [in] four detentions', one former detainee divulged:[31]

> My first detention is Oakington. I stay three days for Oakington. Four months five day in Harmondsworth. After Harmondsworth [name] could bring me to Belmarsh prison – I stay maybe one month. After Belmarsh I stay in Dover detention two months four days. After Dover I go again to Harmondsworth one week. After Harmondsworth I come [name] bring me to Campsfield House...the total's fifteen months in detention.

Although transfers might be prompted by capacity differences at the different centres and a range of other operational reasons (see Bosworth, 2014), various interviewees described the rationale for transfers largely in terms of discipline. 'Troublesome detainees seemed to be moved more', a spiritual support officer reflected.[32] 'Staff frequently say things like "I'll request his removal"; "He's a ringleader, inciting others to cause trouble"', a healthcare worker alleged.[33] This was corroborated by the experience of the detainee who had spent four years in detention: his account of the transfers between centres spanned seven of the eleven detention centres in the United Kingdom, including a high degree of shuttling back and forth, and he was in no doubt that the transfers were often precipitated on disciplinary grounds. 'Sometimes they will accuse me of leading other detainees and things like that' he explained;[34] 'There was one occasion when I complained and another group who followed me and complained. Next thing I was gone from there'.

Some detention centres have better facilities including more hours of free association and better gyms, food, health facilities, libraries, internet provision and relations between officers and detainees. The prospect of being moved to a worse detention centre, or a prison, was therefore upsetting for many of the detainees. I asked one former detainee whether he had been concerned about possible transfers whilst in detention. 'Yes, very worry very', he replied.[35] 'Maybe a manager of staff of detention call you and give you news. Tomorrow you move you go to another detention [where] treatment is different.' 'It used to be Haslow they were threatened with', one activist claimed,[36] 'but now there are worse places. It was Rochester for a time which was an old prison and it was run by prison guards. Asylum seekers had two weeks of it and there they were locked in their cells for 23 hours a day. It was a very brutal place.' As a result of the spatial churning of the detainee population, some detention centres gain a reputation via word of mouth accounts among the detainees. 'Many are moved away and come back again, some two or three times', one healthcare employee reported.[37] This can mean that detainees are grateful for staying put or being returned to certain centres, including Campsfield. I asked one former detainee if people ever return to Campsfield after having been transferred away. 'Yeah yeah, you come back yeah', he told me,[38] 'maybe people enjoy because…Campsfield House is better detention'.

The transfers themselves are disruptive, often breaking detainees' ties to outside support. 'If they get moved to Dungavel [the most northerly major detention centre in the United Kingdom and the furthest from London] and they have a lawyer in London it's hard, the lawyer often gives up the case', one activist told me.[39] Dungavel is in Scotland and the Scottish and English legal systems differ, so this sort of occurrence is not uncommon. Sometimes when detainees are not forewarned about their transfers (which can be a tactic 'in case they are trouble', one staff member claimed,[40] which means that they can be 'informed at 10 pm that they will be moved the next day') then they do not have time to notify their solicitor, especially if they did not happen to have a phonecard or a mobile phone that was working or accessible at the time. It is also more difficult to find people to act as sureties[41] in bail hearings, and can be more difficult to maintain a supply of willing volunteer visitors when the detainees they will be vouching for or visiting are often people that they have met only a few times before, if at all. These side-effects of the transfers left many activists in no doubt about their cruelty. 'People are shuffled around detention centres', the healthcare employee alleged,[42] which acts to

> prevent them getting help to complain about abuse, or with their immigration case. The movements themselves are traumatic and appear to an impartial observer to be part of an intentional system of degrading treatment to prevent people complaining or striving to be accepted as a refugee… People are constantly moved to keep people suppressed and fearful. The staff's understanding shows no doubt about this.

Even within the centre, the transfers put increased strain on the support that detainees were able to access. Staff referred to the considerable challenges posed by a greater turnover, including the difficulty of providing coherent education programmes, the increased demand for health checks and security vetting procedures, and the strain placed upon the centre's ability to cope with short-notice demand for legal help and advice. I was told by staff that there is a 24-hour induction procedure for all new detainees into Campsfield, meaning that during the mid-2000s a significant proportion of the detainees, the so-called 'over-nighters', were never fully inducted. Although the reception suite is well equipped, it is smaller and more sterile than the main facility, and is the site of various tests and procedures, from English language assessment to HIV screening.

One member of the education support team described how her classes (mostly English language classes, but also some computer and arts classes) were disrupted by the turnover:

> We have an incredibly high turnover in class. It's not so difficult for computer classes because everyone works through the course at their own pace. So someone can start any day they like and jog along at their own pace. But for an English class it's impossible to get any sort of cohesion and move along as a group and make any sort of progress. Because you might start with 10 people in your group on a Monday and still have 10 at the end of the week but they're completely different people. The turnover's so high...if they weren't being moved so regularly between centres for seemingly no reason we would have them for a lot longer. [But] detainees will leave us, go to Scotland or London, and they come back to us. So it seems that if they'd stayed with us they could have had an extra month of stability.[43]

The centre manager's actually complained about it', she continued, 'because it's very costly and it's very time consuming. The facilities here were designed as a detention centre to have a much lower throughput. So now it's a higher throughput they're having to deal with many, many more people every day checking them in and out.'

Educational provision was not the only form of support to suffer as a result of the transfers. Medical care was also compromised because many transfers occur overnight. 'A nurse is meant to be present', a healthcare worker pointed out.[44] 'None [are] present at night, but people are then routinely removed at that time' they alleged. One doctor, who had visited many detainees in Campsfield as well as other centres as an independent medical advisor, claimed that detainees' medicines and medical notes would often not be transferred with them from centre to centre. 'They arrive without their medication, they arrive without a medical note', he explained.[45] Subsequently there is a risk that they will not be believed when outlining their need for medication after having been transferred either from another detention centre or a prison. The doctor reported instances in which the induction staff had recorded that detainees 'claimed' to need certain medication. 'Every case which I have seen where it has been recorded that the patient "claimed" to be on some medication or other they bloody well were', the doctor reported, the point being that transfers expose detainees to the risk of not being believed and therefore not having access to the medication they have been prescribed.[46]

A related issue concerns the shortage of interpreters because, as the doctor pointed out, 'many of the people who are in detention do not speak English at all or well'. With a higher 'throughput' of detainees in and out of centres, the demand for interpreters increases because each arriving detainee needs an interpreter as part of their induction. This is especially important in the case of medical examinations that should occur upon arrival.[47] 'If you examine a patient without benefit of an interpreter you are very unlikely to be able to find out the medical facts you need to know to help', the doctor explained. Yet, he continued, 'there is a definite scarcity of interpreters because the use of interpreters costs money. Remember that various centres are run by private companies for profit. With the result that any avoidable expenditure is avoided.'

One questionable solution to this problem is to use same-language detainees as makeshift interpreters. Using same-language detainees in this way 'can be problematic when you are dealing with highly confidential information which might then spread all the way around the detention centre', the doctor explained.[48] In the case of medical information this problem can be particularly acute and put the detainee at risk of 'bullying by other inmates', the healthcare support worker claimed,[49] which often 'goes unreported because people are afraid it will affect their immigration case'. The information that detainees are gay might endanger them, for instance, or at least make it more difficult to settle into a male only centre such as Campsfield. 'Gay people [are] vulnerable [in] the environment', the healthcare support worker claimed; there are 'powerful homophobic group[s] in the centre. Quite a number said they have been harassed.'

In these ways the transfers undermine the support available to the detainees that are moved, and expose them to a series of risks. Moreover, almost inevitably, staff that work in the induction units, in the medical and education teams, as part of the spiritual support of the centre and as regular custody officers begin to view the asylum seekers in their care as a collective rather than as individuals – using terms like 'turnover', 'throughput' and 'deportees' (rather than detainees) to characterise the detention centre's more transient population. Hall (2012, p. 146) found that officers were not keen on detainees 'becoming too friendly' or 'familiar [or] assuming an equal social personhood with the officers', and the staff at Campsfield displayed similar antipathy. The education support worker reflected upon the necessity of remaining detached in the role:[50]

> Some people find it very hard on a personal level too because you hear so
> many horrific things from the detainees about things that have happened to

them or what's going to happen to them if they're sent home. And [often they are in] an entirely hopeless situation because they've no way of proving who they are or what political position they belong to and the danger that they're in. All those things can play on your mind, so you have to develop a sort of professional detachment whereby you listen to them sympathetically at the time and help in any way you can. But when you leave you have to leave that at work. If you were to take it with you I think it would be a huge obstacle because you'd be useless to the next batch that come through.

It is precisely the knowledge that another 'batch' of detainees is about to arrive that makes it imperative to remain professionally detached. The turnover of detainees acts to distract staff from the specificities of individual cases. As she continues,

> It's less difficult now because the turnover's so high. In the past someone would leave and you'd wonder 'what happened to him, I haven't heard from him for months, I wonder if he's still alive'. But now the turnover is so high that the minute someone's left there's someone who's arrived with just as big problems or just such nasty situations, so there's always someone else to help. Nasty as it sounds you very quickly forget the ones who've just left because you're onto the next batch.[51]

For Simmel, the sheer frequency of being confronted by difference is capable of 'stimulat[ing] the nerves to their utmost reactivity until they finally can no longer produce any reaction at all' (Simmel, 1903/2002, p. 14). The result is 'indifference towards the distinctions between [others]' (Simmel, 1903/2002, p. 15) wherein the 'meaning and the value of the distinctions between things, and therewith of the things themselves, are experienced as meaningless. They appear to the blasé person in a homogenous, flat and grey colour with no one of them worthy of being preferred to another' (Simmel, 1903/2002, p. 15). What the spatial churning of detained populations achieves is the more or less mandatory adoption of a detached disposition among staff towards the detainees in their care.

Making Insensitivity Easier

Various other mechanisms operate in Campsfield that not only make insensitivity imperative, but also ease any personal concerns about displaying such detachment. Some staff convince themselves that the detainees are not credible; that they are lying about their cases or that they should not be in the United Kingdom. Another strategy is to belittle the detainees and view their suffering and confusion as trivial and insignificant. The strategy of convincing oneself that the subjects in your care can be treated flippantly can result in the cruel treatment of detainees who are in extremely vulnerable situations. One healthcare support worker, for example, alleged a catalogue of neglectful practices in the mid-2000s, including 'inadequate cleaning', 'lack of advice', 'intimidating drug hunts', 'detention of people who have been tortured [since] immigration do not take it seriously', 'medical treatment withheld', 'hundreds mentally coerced into going [back to their country of origin] every year', 'bullying by other inmates and staff' and 'staff asked to sign forms *after the fact* to say person able to be handcuffed'.[52] These examples of neglect belie a level of insensitivity that is made possible through the belief that the detainees do not possess equivalent standing to staff and consequently can be treated neglectfully legitimately.

There exist institutional prompts that corroborate these self-told ratio-nalisations. One senior charity employee and long-time activist against indefinite detention, outlined the painstaking cataloguing of non-belonging of individual detainees in immigration detention that undermines their credibility and underscores their strangeness.[53] 'The key tactic is to produce migrants as unknowable and anonymous', he suggests with reference to the individual dossiers that are kept on every detainee,

> what's really strange when you read these files and follow cases is govern-
> ments not actually trying to document the person, not trying to prove their
> identity, but building up files to prove their *undocumentability*... Everyone in
> detention gets a document every month setting out the reasons for their
> detention. And it will always, or usually, start name, so and so, aliases and
> there'll be sort of fifteen aliases... When you read it it's actually name,
> Mohamed Ahmed, aliases Ahmed Mohamed, Mohamed with two M's
> Ahmed, Mohamed Ah*mad*. It's just a lot of officials who've spelt his name
> wrong over the years. But he's already being produced as this sinister, unknow-
> able, and criminal, threatening person who requires detention.

In these mundane ways detention centre staff are reminded continuously that the detainees in their care do not belong. 'They...changed my name', one former detainee who had experienced this form of disenfranchisement recalled.[54] 'My name is [first name] [surname]. They changed it to one letter from my first name and one letter from the surname. Then they come back to me and they say "[first name], you are alien. You don't have no right to stay in the country".'

To make matters worse, there are accounts of staff trivialising the experiences and concerns of detainees. Although hunger striking, for example, is an important form of dissent and self-expression in detention (Bousfield, 2005; Edkins and Pin-Fat, 2005; Owens, 2009; Simanowitz, 2010; McGregor, 2011; Conlon, 2013), some respondents told me that staff sometimes did not take food refusal seriously (see Hall, 2012, for similar findings). One health support worker recounted the case of a man who had refused food because he was 'desperate about being sent back and tortured to death as he had been tortured and raped by prison officers in his [home country]'.[55] As a result of his hunger striking the man 'became very weak with his heart not working properly'. At this point he received a visitor but 'staff would not let the visitor see him in the medical ward...presumably because they wanted to punish him for not eating...it took two hours to get him out of bed and into a wheelchair.'

Insufficient respect is also reportedly shown to detainees' possessions. Many items are confiscated upon arrival, which is patronising in itself. '[The detainees are] allowed to take a certain number of changes of clothes and personal items from their baggage', one member of staff explained, 'but the rest of it's then locked away'.[56] One consequence is that when detainees are transferred to different centres they can easily become separated from their effects. 'I lost my stuff', one former detainee told me,[57]

> my big bag, with four or three [pairs of] shoes.

> NG: And did you complain?

> Yeah, I made many, many complaints. I have in my house the letters.

> NG: Were you happy with the response to the complaints?

> Authorities speak to me, write to me but they pay me nothing.

'Some things go to a safe', another former detainee alleged,[58] 'so they put them safe, and [sometimes] they go missing. Like you've got an expensive watch or something, and by the time that people are moved they will look in that safe and they will say "Oh it is not there."' Another consequence of the confiscation of belongings is that, without their belongings, detainees become more reliant upon purchases from the centre shop. Withholding access to the shop, or threatening to do so, then becomes a useful disciplinary tool. One former detainee described being punished in this way for infringing the centre rules:

> You can't spend money, you can spend just one pound or two pounds, and you could lose part or whole of your belongings and that's a torture as well. ... You can buy the food from the shop but they can take it away from you. And it's a common practice, common practice depriving people of their possessions, depriving people of their shopping.[59]

Aside from these sanctions for undesirable behaviour, the rewards that are bestowed on compliant detainees are equally as revealing about the trivialising and trivialised relationship between staff and detainees. One staff member outlined how detainees could earn five pounds for keeping their room tidy,[60] while a recent former detainee divulged that paid work is now allowed by some of the privately contracted security companies within detention. 'You can work for one pound an hour', he told me (this compares to the minimum wage in the United Kingdom of £6.31 per hour at the time of writing),[61] Reflecting on why the rate of pay was so low, the former detainees explained that is was

> because they say 'Well we pay for everything else for you, we paying your house, we're paying your food, so that's what you're gonna get.' Once you are outside you get five pounds an hour and they catch you and say you have no right to work and they take you inside and make you work for one pound an hour, washing, cleaning. For one pound an hour![62]

Paying detainees in such small amounts is more akin to pocket money than a real wage. Not allowing detainees to earn anything like the minimum wage sends a statement to staff that detainees are somehow subjacent to regular citizens. Painting a trivial picture of detainees means that some of their human, adult needs can be overlooked, including their need to know what was going to happen to them next or when their next transfer would be. 'Detainees were given insufficient warning of their next move and were unable to prepare themselves or inform their families, friends or legal representatives what was happening to them', one HMIP report observed (HM Inspectorate of Prisons, 2002). It continued:

> Those granted admission to the UK for the first time were given no help to orientate themselves to life in the UK or to understand the system that would support them [...] Those being removed were lucky to have more than three days' notice, and some did not have that. There was no removal plan which ensured that their affairs in the UK were closed and they knew what to do on arrival at their next destination. No-one would choose to board a plane in these circumstances and it was inappropriate to expect detainees to do so.
> HM Inspectorate of Prisons, 2002, 'An Inspection of Campsfield House Immigration Removal Centre', pp. 12–13

Even the ability to communicate online was curtailed. 'You cannot have a mobile phone with internet access', one detainee explained,[63] 'but you can go to a computer room where you can access material, but only try to connect to what they allow you. Anything else: sorry that's blocked.'

The effect that these mechanisms of control have, not just upon detainees but also upon staff in terms of depicting the detainees in their care as trivial and infantile, impacts upon the way staff treat detainees. Even when staff engage in friendly, supposedly supportive activities, a patronising subtext of superiority and condescension is often discernible. The educator mentioned earlier, for example, emphasised the importance of computer classes not just for educative purposes but because otherwise the detainees are prone to 'just drift in and out and do what they want'.[64] In contrast, the education programme provides structure and purpose and was understood as a way for detainees to demonstrate their strength of character: 'it gives them a chance to prove to you that they have some sort of personal identity', the education coordinator explained,[65] although it was not clear why it might be necessary to do so. As one activist confirmed, the class organisers had a sanctimonious air. 'I can't really explain it but just this, sort of, way you'd make kids do something creative and good for them. They were being friendly but it was very clear that they're working above these people.'[66]

The Cruel Consequences of Insensitivity

In these ways the institutional mechanisms of transfers, sanctions, rewards, dossiers and classes depict detainees fleetingly, trivially and condescendingly, and encourage staff to treat them accordingly. In some cases the sheer unresponsiveness to detainees' suffering illustrates the extent to which they have been trivialised. One member of staff alleged how detainees who were being transferred between centres were sometimes forced to wait an unacceptable amount of time in the security vans that transported them. They 'were not allowed refreshments or to use the toilet for up to 12 hours, with no stops for anything', the staff member claimed, 'people have wet themselves'.[67] In another case an intrusive drug hunt was allegedly allowed to continue with insufficient regard for the psychological implications of the procedure. 'Dogs are brought in every 6 months or so for searches', one health support worker reported.[68] 'A Muslim lad who the dog stopped at [was] taken to his bedroom, asked to take all his clothes off, asked to squat so they could visually check his anus with a nurse and two group four officers. He was traumatized by it, never got over it... Had to go on medication because of it.'

In other cases, the cruel treatment of detainees cannot be explained away in terms of either inattention or an over-fastidious adherence to the rules. For example, at times the desire among staff to avoid disciplinary procedures themselves was apparently valued above detainees' welfare, posing a risk to detainees. One doctor recounted a case in which he visited what he thought to be a torture survivor in detention. Torture survivors should not be detained, but the doctor was concerned that the screening process that was supposed to detect torture survivors was inadequate. What is more, because the detainee had been transferred from a different facility, if the receiving centre found her to be a torture survivor this could implicate the sending centre and potentially land the staff there into trouble. The result, in his opinion, was disregard for clear evidence of torture. 'She showed me two blatant cigarette burns on the back of her right, that is to say dominant, hand', he alleged.[69]

> People do not self-harm with cigarettes on their dominant hand. They do it on the non-dominant hand using the dominant hand if it was self-inflicted. This was a very obvious cigarette burn. It had a...dark edge and it had a diameter of about a little less than a centimetre which is virtually diagnostic for a cigarette burn.

Given the existence of this scar on the dominant hand, the doctor thought the nurse at the receiving centre should have raised the possibility that this person was a torture survivor.[70] What is more, there was little possibility that the scar was missed: 'The nurse had taken the pulse', the doctor explained, 'and had sent a report saying no scars. The pulse was of course immediately adjacent to the wrist, right, that's where you take a pulse.' The only explanation that the doctor could accept was that the nurse had not raised concerns about the detainee being a torture survivor 'because they know that UKBA [United Kingdom Border Agency] do not want a report of torture'.

In this case the anxieties of the nurse around the consequences of admitting that they or someone else had previously made the wrong assessment were allegedly prioritised above the morally correct thing to do. In other cases, the bureaucratic demands of the institution are seem to be valued so much higher than the experiences of the detainee that the immoral consequences of ensuring the smooth running of the system, or breaking down resistance to it, are overlooked. In these instances, the suffering of detainees ceases to matter beyond its ability to frustrate the achievement of institutional goals. Detainees themselves become simply an obstacle to a bureaucratic process; an obstacle that needs to be dismantled. In this vein, activists have reported the abusive use of inter-centre transfers. One activist who had worked with detainees for over 5 years alleged that

> people with serious mental illness [are] being transferred around the detention estate with the explicit purpose of, and you can see on the file, of upsetting them. They are too comfortable in their detention, let's move them somewhere else so that we can upset them and break them so that they agree to move back.

A health support worker alleged the intimidating use of force during removal attempts along similar lines. 'If a detainee fights', she explains,[71]

> the officers scream at them 'f**king keep your hands still or you'll get worse' or shout 'keep your head down', 'get up on your knees'. Some officers obviously love it. The officers have a briefing before removing someone, take off their ties and belts and get worked up for it.

The consequences for some detainees are morally inexcusable. One of my interviewees, a former detainee, recalled his treatment at the hands of private security guards during a deportation attempt. 'If I'm going [back to his country of origin] maybe I will die there. That was why I was not like to go back', he recalled,[72]

> and they tried to force me, and they say they beat me. Then they beat me and then they send me to [detention centre].

NG: When you say that they beat you what do you mean?

When I said I'm not going they take me somewhere, because it was the airport, they took me, put me somewhere, and they started beating me just in the airport…they don't care, they don't care, they just need their job. They need just [that] I go.

NG: Did they use weapons or just hands and feet?

No just with hand they beat me with hand. They beat me, my face. They beat me. They broke my [bone] and when I finished there they send me to [centre].

Such allegations are by no means uncommon. In its study, entitled 'Outsourcing Abuse: the Use and Misuse of State-Sanctioned Force During the Detention and Removal of Asylum Seekers', Medical Justice (2008), a registered charity, reported nearly 300 cases of alleged assaults against asylum deportees, including assaults against pregnant women and children. The allegations ranged from being beaten, choked, kicked, gagged, over-zealously restrained with handcuffs, dragged about, knelt on or sat on to sexual abuse. The alleged consequences ranged from bruising and swelling to fractures, dislocations, cuts, bleeding, and head, neck and back pain. Lord David Ramsbotham, former Chief Inspector of Prisons in England and Wales, wrote in reaction to the report that 'our national reputation is not something to be treated lightly or wantonly, and…if even one of the cases is substantiated, that amounts to something of a preventable national disgrace' (Medical Justice, 2008, p. 1).

Other cruel tactics are apparently motivated by a desire to break down the non-compliance of detainees. 'If you do not get on with them your food will be tampered', one former detainee claimed.[73]

Your room will be messed up, when you go out they can come in and put things in your bed, put rubbish in your bed. They can mess up your food, all things like that… If they don't like you for whatever reason, the colour of your skin, your face, the way your beard grows, you're in trouble.

Conclusion

This chapter has charted the production of a different, up-close, form of indifference to the indifference described in earlier chapters. Where exposure to suffering is frequent there is a possibility that uncalculated compassion and spontaneous kindness could break out and disrupt the smooth functioning of bureaucratic systems of rule that require the morally disinterested treatment of vulnerable individuals. Various institutional features mitigate against this possibility, however, so that compassion is made more costly on the one hand, and insensitivity is made easier on the other. The consequence is that staff are compelled to psychologically avoid their own moral culpability. The ethical command that Levinas describes the suffering other emitting is overruled by the institutional organisation of detention itself, that trivialises detainees.

This is not to suggest that this sort of psychological avoidance of asylum seekers by personnel is absent in other areas of border work such as the back-office decision making, asylum interviews and tribunal appeal hearings discussed in previous chapters. As noted in Chapter Four, staff experience vicarious traumatisation frequently in these areas of work also, so although I have chosen immigration detention as a way to illustrate the effect of psychological avoidance and the means by which it is called forth from functionaries by the institutions within which they work, I do not want to imply that a clean, mutually exclusive separation between moral distance at long range, moral estrangement at close quarters and Simmel's form of indifference always exists between particular sites of immigration control. In reality the situation is more complex and it is possible to identify multiple forms of indifference at various sites.

Conversely, immigration detention is by no means foreign to the sort of long-range moral distance that I set out in Chapters Two and Three. In particular, the separation of decision makers from frontline staff through the mediating effect of the contracting-out of detention centres results in a marked bifurcation of the orchestrators and perpetrators of the detention system in a way that closely resembles the configurations of chains of command that Bauman (1989) associates with moral insensitivity. 'If you spend any time with the detention centre officers, not the immigration officers but the people on the ground in the detention centres', one long-time activist explains,[74]

> they are often...confused [because] you've got two separate sources of authority, you've got the spatialised security institution of the detention centre, which is managed, policed by, controlled by mostly contracted security staff who often don't really know why they are there and what it's all about. And you've got the immigration officers, who are *not there*. ... The immigration officers are in Croydon or Liverpool somewhere. They will never actually visit a detention centre. They never actually meet the person whose case they are managing.

This form of moral distancing corresponds more closely to the first form of indifference outlined in Chapters Two and Three than it does to Hamblet's and Simmel's: the utilisation of literal and organisational distance to occlude the consequences of decisions and to efface the face of suffering others. Yet what immigration detention also achieves – uniquely at least in extent – is a sophisticated use of both the reality and risk of vicarious traumatisation to generate an insensitive workforce. This is achieved through overexposure to suffering, which prompts a reaction in the forms of psychological detachment and avoidance among staff. Such avoidance is made all the more necessary by the frequency of exposure to trauma engendered through the spatial churning of detainees, and it is made easier by the institutional features of detention such as the dossiers, disciplinary systems and education provisions that depict detainees as strange, infantile and trivial. This results in 'compassion avoidance [as] a learned behaviour' (Moeller, 1999, p. 318) that can be, and is, taught by the immigration control system.

We have now arrived at the opposite end of the spectrum of indifference that underpins the various migrant deaths I discussed in the opening section of the book. At this extreme, contact with suffering others itself exacerbates the need for avoidance by threatening to overwhelm empathetic capabilities. Proximity, then, takes on a very different character to that imagined by Bauman, one that is more aligned with Ivan Karamazov's aversion to nearby suffering others (Hamblet, 2003). At this extreme, contact nurtures indifference rather than acting as an antidote to it, as functionaries are systematically exposed to the risk of secondary trauma and must act to protect themselves by 'renouncing the response' (Simmel, 1903/2002, p. 14) that comes most naturally when exposed to emotional stimuli – in proportion to the contact they experience. At this extreme, the relationship between 'distance' and morality is also complicated and distorted. Bauman's thesis that the closing of physical distance increases morality does not apply here: rather a more complex, two-step, process of over-closeness that precipitates psychological distancing is in evidence.

The result is that in immigration detention we discover the limits of Bauman's ethics of the face to face encounter that values individuals in their specificity and uniqueness and usurps depersonalised structures of control. Admittedly there is always the potential for empathy as Hall (2012) has noted. Nevertheless the form of indifference identified here constitutes a powerful ancillary consequence of the organisation and administration of border control that, in its fullest form, is perfectly capable of drawing forth the sort of cruelty and degrading treatment that the abused men in Harmondsworth Detention Centre in 2013 were subjected to.

Notes

1   Not just government employees, but also asylum support group workers in the United States and the United Kingdom regularly suffer compassion fatigue that can lead them to 'become jaded, cynical, brusque, not really having the patience to go to the meetings or…explain why asylum seekers aren't the same as terrorists' (Gill *et al.*, 2012, p. 23). Some of the issues facing immigration detention centre staff, including how secondary trauma can impact upon the personal lives of staff, are the subject of an innovative illustration of an

account of one guard's experience of working in an immigration detention centre in Australia – see Olle *et al.*, 2013.

2   Interview with asylum seeker and former detainee, 2006.

3   Interview with activist and detainee visitor, 29 May 2006.

4   An independent consultant, Matrix Evidence, that provides advisory services to public and private sector organisations, foundations and charities, found that by making simple improvements to the assessment of the likelihood of eventual deportation and return at the start of a detainee's case and releasing those not expected to be deportable, the United Kingdom stands to save over £344 million, more than double the monitoring cost of ensuring that every detainee who was consequently released did not abscond (Marsh *et al.*, 2012).

5   As the charity Liberty (2014) explains, '[a] decision to detain is made by individual immigration officers and, unlike people detained under the criminal justice system, its lawfulness is not automatically subject to independent review. A detained person can, after seven days have passed, apply to a judge for review of his or her detention, but many people, particularly those who don't speak English, are unaware of this procedure and find it difficult to access legal advice.'

6   These data are taken from the following website: http://www.no-deportations. org.uk/Media-1-2012/Self-Harm2013.html (accessed 19 November 2015).

7   Interview with activist and visitor to immigration detention, 19 May 2006.

8   This evidence is taken from a collection of accounts of immigration detention in the United Kingdom and Australia published by Barbed Wire Britain in 2006, entitled 'Voices From Detention II' (Garcia *et al.*, 2006, p. 32). The individual quoted won his case after two years and has been recognised as a refugee under the 1951 Geneva Convention.

9   Interview with staff member working in immigration detention, 14 July 2006.

10   Interview with staff member in immigration detention, 14 July 2006.

11   Interview with spiritual support team member, July 2006.

12   Interview with former detainee, 6 February 2014.

13   This evidence is taken from a collection of accounts of immigration detention in the United Kingdom and Australia published by Barbed Wire Britain in 2006, entitled 'Voices From Detention II' (Garcia *et al.*, 2006, p. 30). The individual quoted was forced to flee from Nigeria.

14   Interview with staff member in immigration detention, 14 July 2006.

15   Interview with activist and detainee visitor, 29 May 2006.

16   Interview with activist and detainee visitor, 29 May 2006.

17   Interview with staff member in immigration detention, 14 July 2006.

18   This evidence is taken from a collection of accounts of immigration detention in the UK and Australia published by Barbed Wire Britain in 2006, entitled 'Voices From Detention II' (Garcia *et al.*, 2006, p. 32). The individual quoted was forced to flee from Nigeria.

19   A pseudonym.

20   Interview with activist and detainee visitor, 23 May 2006.

21   Interview with activist and detainee visitor, 29 May 2006.

22   Interview with staff member in immigration detention, 14 July 2006.

23   Interview with former detainee, 14 July 2006.

24   Interview with former detainee, 14 July 2006.

25 Interview with member of immigration detention centre support staff, December 2005.
26 Interview with detention centre spiritual support team member, July 2007.
27 Interview with former detainee, 6 February 2014.
28 Interview with welfare team member in immigration detention, 3 July 2006.
29 Interview with educator in immigration detention, July 2006.
30 Interview with activist and visitor to immigration detainees, 19 May 2006.
31 Interview with former detainee, mid-2000s.
32 Interview with staff member in immigration detention, July 2006.
33 Written communication from healthcare worker in immigration detention, March 2005.
34 Interview with former detainee, 6 February 2014.
35 Interview with former detainee, July 2006.
36 Interview with activist, 28 May 2006.
37 Written communication from healthcare worker in immigration detention, March 2005.
38 Interview with former detainee, July 2006.
39 Interview with activist, 28 May 2006.
40 Written communication from healthcare worker in immigration detention, March 2005.
41 Sureties pledge a certain amount of money at the detainee's bail hearing that they will pay to the government if the detainee absconds.
42 Written communication from healthcare worker in immigration detention, March 2005.
43 Interview with educator in immigration detention, July 2006.
44 Written communication from healthcare worker in immigration detention, March 2005.
45 Interview with doctor who has visited immigration detainees, October 2010.
46 Breaks in HIV and TB treatment in particular can be extremely dangerous because they can allow disease progression.
47 I asked one doctor who had visited many detainees in various detention centres in the United Kingdom if there are any questionnaires that detainees can complete in their own language about their health as a way to screen for medical issues upon arrival. 'There are', the doctor replied, 'and there are translation sheets which are adequate I suppose for a quick screen but they are not adequate for a medical examination'.
48 Interview with doctor who has visited immigration detainees, October 2010.
49 Written communication from healthcare worker in immigration detention, March 2005.
50 Interview with educator in immigration detention, July 2006.
51 Interview with educator in immigration detention, July 2006.
52 Written communication from healthcare worker in immigration detention, March 2005.
53 The charity worker was talking at an ESRC-funded seminar on immigration detention.
54 Interview with former detainee, 6 February 2014.
55 Written communication from healthcare worker in immigration detention, March 2005.

56   Interview with educator in immigration detention, July 2006.
57   Interview with former detainee, July 2006.
58   Interview with former detainee, 6 February 2014.
59   Interview with former detainee, February 2014.
60   It may be no coincidence that phonecards – which represent some detainees'
      only link to the outside world – at the time also cost five pounds
61   Interview with former detainee, 6 February 2014.
62   The *Guardian* ran a story in 2014 that confirmed these practices (Rawlinson,
      2014).
63   Interview with former detainee, 6 February 2014.
64   Interview with educator in immigration detention, July 2006.
65   Interview with educator in immigration detention, July 2006.
66   Interview with activist and detainee visitor, 29 May 2006.
67   Written communication from healthcare worker in immigration detention,
      March 2005.
68   Written communication from healthcare worker in immigration detention,
      March 2005.
69   Interview with doctor who has visited immigration detainees, October 2010.
70   Rule 35 of the Detention Centre Rules (2001) requires detention centre
      medical staff to report anyone for whom detention is harmful or who may have
      been a victim of torture. However, the Independent Chief Inspector of Borders
      and Immigration and HM Inspectorate of Prisons (2012) have criticised the
      poor quality of these reports and of the responses that Home Office case owners
      have made to them.
71   Written communication from healthcare worker in immigration detention,
      March 2005.
72   Interview with asylum seeker and former detainee, mid-2000s.
73   Interview with former detainee, 6 February 2014.
74   The activist was talking at an ESRC-funded seminar on immigration detention.

Chapter Six
Indifference and Emotions

Violence has been turned into a technique. Like all techniques, it is free from emotions and purely rational.

Bauman, 1989, p. 98

I have now set out three critical developments of the moral distance argument I presented in Chapter Two, arguing that moral distancing of subjects is only part of the story, albeit an important part, in accounting for the indifference that pervades and facilitates border control. In making this argument I have employed the ideas of a set of theorists of indifference, bureaucracy and moral distance – including Bauman, Simmel, Glover and Weber – in exposing the diversity of distancing mechanisms and forms of indifference in evidence throughout Britain's border control industry. In this chapter, I challenge their approach in an important respect. Whereas they see estrangement and moral indifference as essentially devoid of emotions, I show that emotions are compatible with, and sometimes central to, the mechanisms through which indifference is achieved.

My focus on emotions in this chapter chimes closely with the burgeoning interest in emotions among human geographers in recent years, both in general terms and with specific reference to policy makers. Geographers have reacted strongly against formal, supposedly emotionless, conceptions of space that, for Henri Lefebvre at least, have become increasingly dominant in contemporary society. Lefebvre identifies the rise of 'abstract'

Nothing Personal?: Geographies of Governing and Activism in the British Asylum System, First Edition. Nick Gill.

conceptions of space that are associated with 'order' (Lefebvre, 1991, p. 33) such as those typically used by 'scientists, planners, urbanists, technocratic subdividers and social engineers' (Lefebvre, 1991, p. 38). This type of space 'claims to be neutral, universal, apolitical, value and emotion free' (Smith *et al.*, 2009, p. 2) but in fact tends to reduce the complexity of human life into a stripped down plan, the pursuit and realisation of which subsequently results in the sidelining of the everyday stuff of life itself.

Against these 'lethal' conceptions of abstract space (Lefebvre, 1991, p. 370) Lefebvre identifies the importance of lived space, which refers to the open space of the everyday: a 'concrete ... which is to say, subjective' space (Lefebvre, 1991, p. 42). Lived space embraces the 'loci of passion, of action and of lived situations' (Lefebvre, 1991, p. 42). Geographers who have attended to emotions in recent years have often done so in order to provide an antidote to abstract space by focusing on the everyday, lived happenings of social life that they refuse to proceduralise, flatten and understand as an ancilliary part of a wider rational scheme. Their interest is prompted by a desire to resist the type of abstract spatial conceptions that 'facilitat[e] political, bureaucratic, and technological interventions of a kind that regard emotional involvements as ... faults to be corrected' (Smith *et al.*, 2009, p. 2).

The upshot has been a sustained interest in both the 'incandescent passions' that occasionally make their mark, as well as the less intense 'constant swirl of emotions that takes us through the day' (Amin and Thrift, 2002, p. 83; see also Davidson *et al.*, 2008). Emotional geographies has become a vibrant subfield.[1] One field of interest has been in the emotional experience of policy making and governance (Hunter, 2012). Horton and Kraftl (2009), for example, discern 'fundamental ... connections between that field of practice conventionally known as policy and that range of affective, bodily intensities conventionally named emotion' (Horton and Kraftl, 2009, p. 2985). In spite of a lingering belief among policy makers that emotional considerations are '"detached from [the] real world"'(Horton and Kraftl, 2009, p. 2986, citing Hamnett, 1997, p. 127), and a persistent perception among some geographers and other social scientists interested in emotions that policy analysis is 'dull, lumpen [and] atheoretical' (Horton and Kraftl, 2009, p. 2986), work on the emotions of policy makers and decision makers has aimed to illuminate the 'enlivened' (Smith *et al.*, 2010) spaces of policy practice within institutional settings. The discussion in this chapter contributes to this work.

My first argument is that moral indifference in bureaucracies often arises as a result of an *emotionally conflicted* state wherein empathetic compassion is overridden by a variety of other concerns. Following Nussbaum (2001), this approach holds that there are powerful emotional impediments to compassion. This insight is important not only because the approach employed by the theorists of moral distance that I listed above tends to imply that moral distance is generally emotion-free, but also because where they do recognise the operation of emotional influences over morality, these are invariably seen to act in order to soften rather than harden attitudes towards suffering others. Both assumptions, I argue, are incorrect.

My second argument focuses in particular upon feelings of anxiety among border control functionaries. Although it is possible to discern the influence of a series of different types of emotions over empathetic compassion in British immigration control, it is anxiety, I argue, that constitutes the most pervasive check over moral sensitivity in this arena. Anxiety among functionaries is largely traceable to the influence of the press in Britain, which has had the effect of making workers fearful of the consequences of acting humanely beyond (and sometimes even within) the terms of their employment. By making this argument I depart slightly from Nussbaum's approach, as she identifies both shame and envy as important impediments over compassion, neither of which revealed themselves as particularly important in constraining compassion in the course of my own investigations. Indeed, where I came across evidence of shame among functionaries in particular, this was more likely to be associated with an increasingly sensitive approach to their work, rather than a less sensitive one (my argument more closely approximates the position of Ahmed, 2004, in this respect). Nevertheless, I retain Nussbaum's basic argument to the extent that compassion among functionaries appeared to be trumped by various counter-posed emotions, with anxiety figuring prominently amongst them.

An important implication of this insight is that it is necessary to draw a distinction between Bauman's notions of insensitivity and indifference (which he tends to use more or less interchangeably and which, up to this point, I have done also). Insensitivity implies a lack of internal emotional experience, an imperviousness that sits comfortably with the eliding of moral distance and emotionlessness that I am objecting to. Indifference, on the other hand, might well involve passionately felt, yet counter-posed, feelings. A prisoner, for example, could be indifferent between the prospect of five years more confinement in a regular prison or two more years of hard physical demands in a boot camp, but care deeply about both options. According to my findings, functionaries of border controls tend to be more likely to be indifferent than insensitive to the plight of suffering others, even though the upshot of both emotional conditions is often comparable in terms of their lack of compassion.

My third argument concerns the contradictory functions of a set of softer emotions including care, empathy and compassion in the composition of indifference. An ethic of care has been vaunted as an antidote to dry, rule-based forms of justice (Tronto, 1993). But a concerning element of the immigration control system is the way in which care features as an element of control and subjugation. Care and indifference have become increasingly closely intertwined, I argue, and, at its worst, care can now be seen to operate as a screen for morally dubious actions. This argument prompts us to question whether these softer emotions are really a good thing in the context of border control – a theme that I maintain in the next chapter about compassion.

Overall, the chapter demonstrates the emotional character of much of everyday life in immigration control, as well as the connections between emotions and the way that policies are implemented, by drawing on a series of examples from across the sites and institutions of control already discussed in the previous chapters. In so doing the chapter identifies the role played by emotions in moral indifference as an important, and often overlooked, consideration.

Emotions Versus Indifference?

The theorists I have mentioned associate insensitivity and indifference with emotionlessness in various ways. Both Glover and Simmel, for example, emphasise how the conditions of contemporary social life have undermined the supposed pre-modern equilibrium between emotions and morality. For Glover, a primary challenge of globalisation is that it drives a wedge between the people whose lives we can profoundly affect, and the people we feel emotional about. According to this argument, the extent of our influence over others is now much wider than in previous eras due to the effects of modern technological advances and the global market economy. But our emotional responses privilege only the near at hand. Hence even when we have a strongly held conviction about the morally correct thing to do regarding the distant needy, Glover argues, '[o]ur beliefs start to diverge from the emotional responses natural to us' (Glover, 1977, p. 294). This is enough to make doing the right thing – such as giving significantly more money to international charities, for example – very difficult indeed. Or, put more precisely, our lack of emotional responses to the distant needy is likely to make doing what we might believe to be the wrong thing when pressed – such as *not* donating to international charities – much easier.

For his part Simmel (1903/2002), in describing the pitfalls of modern cities, identifies the emergence of an implacable, emotionless metropolitan type of character, as noted in the previous chapter. 'Instead of reacting emotionally' (Simmel, 1903/2002, p. 12) to the variety of people and situations that they encounter in urban settings, Simmel writes, people with this type of character react 'primarily in a rational manner' (Simmel, 1903/2002, p. 12) that can be characterised by 'a purely matter-of-fact attitude in the treatment of persons and things in which a formal justice is often combined with an unrelenting hardness' (Simmel, 1903/2002, p. 12). He continues:

> The purely intellectualistic person is indifferent to all things personal because, out of them, relationships and reactions develop which are not to be completely understood by purely rational methods. ... These relationships stand in distinct contrast with the nature of the smaller circle in which the inevitable knowledge of individual characteristics produces, with an equal inevitability, an emotional tone in conduct, a sphere which is beyond the mere objective weighting of tasks performed and payments made.
>
> Simmel, 1903/2002, p. 12

Like Glover, Simmel employs a reading of the past that, perhaps naïvely, envisages an historical, pre-modern (and in Simmel's case rural) spatial equilibrium between the emotional capacities of individuals and the reach, frequency and intensity of their activities (Amin and Thrift, 2002, for instance, are strongly critical of the nostalgia they detect in Simmel's work). Both Glover and Simmel also explicitly contrast emotions and personalism with hardness, rationality and indifference.

Weber and Bauman's work on bureaucracy shares the same basic assumption as Glover and Simmel that emotions and indifference are opposed. Weber was least interested in what he considered the lowest form of action, which he called affectual action, defined as action based upon individual emotions in specific situations (Macon, 2012). He argued that the more sophisticated bureaucracies become the less common this type of action is. 'The more perfectly the bureaucracy is dehumanized,' Weber (1922, p. 15) writes, 'the more completely it succeeds in eliminating from official business love, hatred, and purely personal, irrational and emotional elements which escape calculation.' The natural result of bureaucratic systems is therefore the dominance of 'a spirit of formalistic impersonality' that can be characterised as 'without hatred or passion, and hence without affection or enthusiasm' (Weber, 1922, pp. 15–16). 'This is the spirit in which an official conducts his office', Weber argued (Weber, 1922, pp. 15–16), '[o]therwise the door would be open to arbitrariness'.

Bauman similarly associates the disastrous moral estrangement and indifference that facilitated the Holocaust with a lack of emotions, noting the surprising lack of violence that accompanied systematic mass murder. '[T]here was not enough "mob" to be violent', he observes (Bauman, 1989, p. 74), '[m]ass destruction was accompanied not by the uproar of emotions, but the dead silence of unconcern. It was not public rejoicing, but public indifference which [facilitated the Holocaust].' 'Stalin's and Hitler's victims', he continues (Bauman, 1989, p. 92), 'were killed in a dull, mechanical fashion with no human emotions ... to enliven it.'

We can raise a series of objections to the way in which emotions and indifference are separated by these authors. Emotions are dealt with as a whole as if this constitutes a coherent category, rather than by giving attention to specific emotions that may very well have different tendencies and properties to each other and behave in very different ways (Henderson, 2008). Furthermore, the process of estrangement itself is assumed to be largely impassive and, where emotions are discussed, these are seen as detrimental to moral insensitivity. Neither of these assumptions conforms to the evidence provided by my interviewees, who experienced a range of specific emotions that ran alongside, and often aided, the process of distancing and the nurturing of indifference. In the next sections I show that a variety of specific emotions can promote moral estrangement and indifference.

The Interplay of Emotions in Immigration Control

Admittedly, I occasionally uncovered evidence of emotions that worked against the process of moral estrangement, as these key theorists assume. Shame, for example, was sometimes discernible. 'One of the men working on the security door on the way into Croydon was actually giving up his job that week', one activist who had recently visited Lunar House told me,[2] 'because he couldn't live with himself any more treating people the way he was expected to treat people as part of his job'. Another immigration officer, whom I met at the counter-protest against the EVF [English Volunteer Force] outside Lunar House, recalled his work in the removals section of the Home Office, including his role in deporting children: 'I hated that. Hated it. Really nasty work', he told me.[3] 'Eventually I moved from Liverpool to Croydon and took on different work, regular casework.' Another immigration officer who was temporarily involved in a refugee resettlement scheme run by the Home Office expressed his relief at being asked to do resettlement work, which revealed his view of the work that he usually did and that his colleagues continued to be engaged in. 'I am one of the lucky ones', he reflected,[4] 'because I get to do some of the good work with refugee communities.'

Shame seemed particularly likely to be experienced by frontline staff who were expected to maintain prolonged relationships with migrants in difficult situations. 'I see them in the coffee shop and they come up and they're bloody starving', recalled one police officer whose regular patrol included an area where many destitute asylum seekers resided.[5] 'How are they going to feed themselves', he continued,

> unless they're gonna find an employer that's going to exploit them totally, pay them peanuts for doing long hours they have to feed themselves by resorting to crime, petty crime. So there's an immediate effect with the police and society in general. And it propagates this popular myth that the popular press likes to propagate that, you know, asylum seekers are taking our money, that they're committing crime. Of course they're committing crime because they've got no way of bloody feeding themselves! And so police time gets

taken up, costs a lot of money to get them in the court system, identifying them costs more money. It costs far more money in the long run to deal with these people than it does to support them. The whole thing just doesn't make sense at all.

This sort of critical awareness of the failure of the system of border control from exasperated insiders has long been recognised as an important and potentially potent social phenomenon (Goffman, 1967; Scheff, 1988). In certain circumstances, for example, it can produce compassion beyond the formal demands of the employee's role. One activist recounted a situation in which a suicidally depressed child had been released from hospital into police custody as a result of an administrative error. 'The police officers were very sympathetic', the activist recalled,[6]

> because they were horrified they had to do this to a frightened little girl. One police officer had a daughter that looked the same age as her and he was ashamed that he actually had to keep this kid in a locked cell overnight. She [the asylum seeking girl] said that he kept going in during the night to check that she was alright and to see if she wanted anything to eat because she was curled up in the foetal position in the corner of the cell.

The link between compassion and shame has meant that shame has been viewed by some thinkers as 'crucial to moral development' (Ahmed, 2004, p. 106). Admitting failure to live up to an ideal such as treating suffering others fairly and kindly 'is a way of taking up that ideal and confirming its necessity' (Ahmed, 2004, p. 106). 'Despite the negation of shame experiences,' Ahmed explains, 'shame confirms ... commitment to such ideals in the first place' (Ahmed, 2004, p. 106). In this vein shame has been taken to indicate the existence of a 'lay morality' that is impervious to institutional influences (Sayer, 2005, p. 948). Where shame involves an uncomfortable dissonance between personal ethics and the demands of institutions there may even be scope for insider-led reform.

Unfortunately, however, anxiety was a more powerful and more common emotion among my interviewees and tended to pave the way for morally insensitive systems of control. Although there was some evidence of anxiety over disciplinary procedures among staff who had made mistakes (e.g. as noted in the previous chapter, one visiting doctor to immigration detainees claimed that detainee health problems are not always accurately reported because staff are afraid of getting 'in trouble' and are 'frightened of the consequence'), the central source of anxiety was traceable to the right-wing printed media in the United Kingdom. 'Every time there is a scaremongering head line in the *Daily Mail* or the *Daily Mirror* or *The Sun* about immigration', one activist told me,[7] 'that translates directly into pressure inside Lunar House and throughout the IND [Immigration and Nationality Directorate]. There's a kind of paranoia that it produces and the anxieties about the next scandal that is going to come out or be leaked.' The British printed newspaper press have a track record of naming and shaming not only individual employees who have acted incompetently, but also those who are deemed to have acted too liberally and facilitated the entry of too many migrants into the United Kingdom. In 2012, for example, *The Telegraph* website carried a story that listed the names of three senior judges who 'rule far more often than others in favour of offenders seeking to avoid deportation' (Barrett and Ensor, 2012). Despite the fact that these judges may very well have received certain types of cases that were more likely to warrant certain rulings, and had in any case operated entirely within the law and the boundaries of their judicial discretion, this sort of publicity runs the risk of dampening willingness to act courageously and independently in the exercise of immigration powers in the future.

Given the ever-present risk of negative media attention, ensuring that the task of managing the media image of the IND was properly resourced was actually a central concern of senior management. 'We don't have our own press office, that's in the central Home Office and that provides a challenge', one senior manager at Lunar House told me.[8] 'You have to make sure that we're actually going to be joined up … because we do get hit by the press'. Negative media coverage of government departments poses risks not only in terms of popular perceptions of a state failing to effectively control migration, but also for staff morale, which has implications for productivity and staff turnover. The manager seemed at least as concerned about negative press coverage from this perspective. 'What's happening is people [referring to her own staff in the IND] are seeing the press and we need to have a line to staff as well', she explained;

> We need to get them to understand … because staff pick up stuff in the press and think "Oh, is that true? I didn't know about that". Actually most of the time it isn't true, but there is something in it, there is a germ in it, and we need to explain to staff that this is where this is from and this is the background to it and this is what we're doing about it etc., etc. So that's quite important[9]

With these risks in mind, the imperative to avoid situations that might lead to negative media coverage can sometimes make management extremely wary of forging working relationships with unknown groups. When South London Citizens (SLC) approached the senior management team to conduct their investigation into how to improve customer service at Lunar House, for example, the management's initial response was guarded. 'I mean the danger for us is that there will always be a journalist who will pick this up in six months' time and think "ooh" or whatever', explained one of the management team.[10] It was only when the activists were able to demonstrate that, in the words of one senior manager, 'they were very well meaning, their hearts were in the right place [and] they didn't want to have a pop at IND' that the management team were prepared to work with them.[11]

The anxiety that SLC provoked also led to some bizarre behaviour, indicating the influence that trepidation over media coverage can have. In one notable instance, the queue outside Lunar House was allegedly 'hidden' in anticipation of negative media attention. According to a number of SLC activists, on the day that they arrived at Lunar House in order to give tea and coffee to asylum seekers and other migrants waiting in the queue, the queue itself began in the small hours of the morning. At 9 am, however, the waiting people were beckoned inside the building so that, by the time the activists arrived at 10 am, there was only a fraction of the number that had been there only an hour before. As one local newspaper reported, quoting an activist who was involved: 'The queue had magically disappeared. There were a few people queuing at the back of the building but the majority of them had mysteriously gone. We believe the Home Office deliberately hid them to play down our concerns' (McQueenie, 2005). 'What the IND did', one activist that I interviewed asserted,[12] 'thinking that they'd be faced with some really unpleasant public scene, was they just moved everyone inside, so they basically put people up staircases and things to make sure that there was no queue'.

This is not to say that the effect of the threat of negative media coverage is confined to high-profile centres of control like Lunar House. Employees never know when the next media storm will strike, or to which part of the system of border control it will refer. This uncertainty over the precise nature of the risk that is faced is a key characteristic of anxiety. Whereas fear can be described as an emotional reaction 'to a threat that is identifiable' (Rachman, 1998, pp. 2–3; see also Fischer, 1970), anxiety describes the 'tense anticipation of a threatening but vague event' or a feeling of 'uneasy suspense' (Rachman, 1998, pp. 2–3). The upshot is that the response to anxiety is generalised rather than specific – everyone within the bureaucracy must take the necessary precautions.

Consequently, when an individual receives negative media attention whole tranches of employees in similar or related roles can be put on alert. When one immigration judge made some unfortunate comments[13] about his work on a radio programme in 2010, for example, other immigration judges were advised to be cautious, including a recent interviewee of mine. 'Please excuse this brief informal note following the interview you conducted with me about my judicial work in the asylum and immigration field', wrote my interviewee, who decided to contact me directly about the incident;[14]

> Since then there has been an incident when one of the judges ill-advisedly participated in a Radio 5 live 'phone in' programme making therein some unguarded and unfortunate comments not thought to reflect well upon the

judiciary. Since the programme, individual judges have been advised to steer clear of the media ... There is therefore some reluctance by judges to get involved in any kind of discussion about their work without first getting approval. ... I have not spoken to the judges to whom I referred when we met [my interviewee had offered to help recruit other judges for my research], but others to whom I have spoken have been somewhat cautious and apprehensive since the radio programme about taking part in any interviews at all.

In fact, the process of contacting and recruiting research participants for my research was revealing in other ways. Around three times as many people in official positions that I approached for an interview refused to participate in my research than agreed, for example – either explicitly or, more commonly, through resolutely ignoring my requests. Where refusals were explicit the risk of negative media attention was the usual justification. This cautious approach of my interviewees also extended to the interviews themselves. Interviewees would frequently reserve their richest reflections for the period immediately after the voice recorder was turned off (much to my frustration), usually citing the desire to talk to me 'off the record'. They often also wanted to know precisely what would happen to the transcripts, and sometimes checked over them thoroughly. Some officials warned about what could happen to my thesis (some of the research was doctoral research) if anything was misquoted or misrepresented, and some would talk darkly about work being 'impounded' in the interests of security, even though the feasibility of this was never made clear to me.

Compounding this sense of caution was the fact that many of the people I invited to take part in the research were alarmed if I adopted anything approaching a formal manner. Asking government employees to sign consent forms, for example, often induced suspicion and reluctance. And dressing smartly, as is the usual way to approach 'elite' interviews (Valentine, 1997), also seemed to disturb their image of me as a harmless researcher and I soon gave up the practice in favour of informal dress, which was apparently much less threatening.

One consequence of media-related anxiety is that organisations that have the potential to cast government departments in a positive or negative light in the press can command a high degree of leverage among immigration sector decision makers. One senior manager, for example, was appreciative of SLC's willingness to cushion the impact of the 'sex for visas' scandal (mentioned in Chapter One) that broke in the mid-2000s (Doward and Townsend, 2006).[15] 'We had a ... scandal in January when there was these accusations that people in the public enquiry office had treated certain applicants more fairly and demanded sexual favours and stuff like that', she recalled,[16]

> but the press went to [the SLC activists] for a quote and actually it was quite good that we've got a good relationship because they said 'No actually, we didn't see any of that, they need to do a lot about waiting times and comfort of

the building and maybe change the facilities and blah blah blah, but actually the integrity of the caseworkers we wouldn't question'. And that's actually really useful for us because we could have said those things, whereas actually it's not that effective when you say it yourself ... but if one of the stakeholders does, it's actually quite effective.

It is important to note that many migrant support and activist groups make the decision not to have any dealings with the press for very good reasons. Some worry that press involvement around particular migrants' cases might backfire and incriminate or endanger them in some way. 'I mean their families could be targeted back in their home countries', one activist explained.[17] Others discuss the risk that immigration decision makers like caseworkers and judges who feel they are being pressurised into making a particular decision on a case by featuring the case in the media might respond angrily and make the opposite decision. Others fear that embarrassing the government is 'risking some kind of retribution. There's been stories where [a government department] has actually gone back and tried to dig up additional information on someone because they were on the front page of the newspaper to try to discredit that person', one activist noted.[18] For many activists, advocates and charity workers these risks are enough for them to decide that they want no dealings with the press whatsoever. '[I]t seems to me the only thing the press could do is screw me up', one legal advocate explained.[19] 'I can't be putting my services in the way of any kind of danger just for the sake of issuing an interest story', another noted,[20] 'I just don't see the benefit of it'. Some activists even felt that they were more likely to be able to influence government policy or practice when government departments were out of the limelight. 'You are more likely to have a bit of space to operate in with government off the front pages,' one advocate for refugee issues suggested,[21] 'because they don't feel so pressurised.'

Other activist groups, however, see anxiety among decision makers about negative press coverage as potentially exploitable. After the queue-hiding incident, for example, the SLC activists at Lunar House were able to capitalise upon the anxiety about further negative media coverage among the management team to elicit their cooperation in implementing a series of reforms throughout the building. Although the management team had grudgingly agreed to allow their investigation to take place and cooperate where necessary after the launch, according to various SLC activists they would routinely postpone meetings, withhold information, miss deadlines, attempt to cancel appointments and leave very long periods of time between correspondence. The activists responded by threatening another eye-catching, media-attracting public action, such as a parade or distributing more tea and coffee at Lunar House. As one organiser explained, reflecting upon the process of securing the cooperation of the senior IND management team during this time, what the managers longed for above all was the absence of media. If this was assured, then everything else seemed possible:

> [After the initial distribution of tea and coffee] they were taking us very seriously because they saw what we'd done, we'd got quite a bit of media attention. When we heard from [senior immigration officials] 'Yes, we'll come to discuss working together on the basis that there is no media', we said 'OK, we agree'. And when they hadn't sent us their response ... we decided that we would stage an action, if only to galvanise the support of the voluntary sector. But having let them know what we were planning to do, we then got an immediate response back.

Having received this response SLC then called off their action. The same activist recalled that '[the IND] were so relieved that we'd called it off that [name] the [senior position within the IND] came over to the church where we meet to thank us and to talk to us about the response that they'd sent'.[22]

There is, then, a complex interplay of emotions that both challenge and promote moral indifference. Border control employees sometimes feel ashamed about their role in systems that they can see are immoral. But more commonly they feel anxious: about their jobs, about disciplinary measures and about the next media storm. Anxiety about the tabloid printed press in particular has the effect of disciplining them by making them more compliant to sources of pressure that either threaten them with the risk of unwanted media attention or promise to reduce this risk. It seems clear that this interplay resembles neither the emotionlessness that Glover, Simmel, Bauman and Weber associate with morally indifferent bureaucracies, nor the opposition between emotions and moral indifference that they posit.

The Softer Side of British Immigration Control

Alongside shame and anxiety, the rhetoric of international immigration control now regularly cites softer emotions like care, empathy and compassion. Here I focus on the claims made by immigration control organisations that they are actively pursuing these emotions among their functionaries and throughout their cultures. Bialasiewicz (2012), for example, notes how the renaming of detention centres to 'care centres' in Libya allowed Libyan officials to claim that the centres, which were 'often the site of forced labour, rapes and beatings' (Bialasiewicz, 2012, p. 855), could ironically now be justified as attempts 'to ensure the physical integrity of irregular migrants, particularly as they expose themselves to dangerous situations' (Bialasiewicz, 2012, p. 855, quoting one of the Libyan officials). The claim is that these repressive centres somehow have the best interests of migrants in mind.[23]

This same claim recurs in the British case and in what follows I explore how this emotional apparatus is rolled out alongside and in the service of moral indifference. One senior manager talked passionately about the values of the Home Office when I interviewed her in the mid-2000s. 'We've recently introduced some values', she explained,[24]

> and we want to make sure we embed those values in terms of not just having nice posters on the walls. The values are: we will be professional, innovative, open, collaborative, having respect for people, nothing surprising in there, but frankly it's not going to have any effect unless we actually try to translate that into behaviours.

Since their inception, these and similar values have been rolled out to underpin many of the seemingly brutal elements of immigration control. Job advertisements to recruit detention custody officers, for example, which include 'day-in-the-life' style accounts from existing employees, are issued to job centres and newspapers across the United Kingdom and strongly emphasise the need for empathetic staff. 'The best people in this job have excellent communication skills, are willing to be flexible and most importantly be able to empathise with others', one advert reads (G4S [Group 4 Securicor], 2013a). 'The secret to this job is all about talking to people. It's about speaking to people and treating them with respect', it continues. Key skills include the ability to listen to others and be non-judgemental.

> Some of the detainees we escort are under a great deal of stress and are often confused or scared, so the ability to understand and deal calmly with all kinds of people is essential. We always have to be observant and aware as well as patient when talking to detainees too. But the most important thing is to be non-judgemental and treat everyone as you would like to be treated yourself.
>
> G4S, 2013a

In another example, an existing overseas escorting officer describes his actions during a deportation procedure. 'I always ask how the detainee is feeling about the removal', he tells us, and whilst on the plane '[w]e maintain a constant stream of conversation' with the aim of 'establishing a rapport' (G4S, 2013b). Similarly the Border Agency's guidelines for training detention custody officers[25] stresses that the duties of a custody officer include 'attend[ing] to a detainee's well-being' alongside their duties to prevent escape and maintain discipline (UK Border Agency, 2011).

Elsewhere, one of the most controversial extensions to the British detention estate fully embraces the softer side of immigration control. In 2011 a new specialist family detention centre named Cedars – an acronym for compassion, empathy, dignity, approachability, respect and support – was announced. The centre was to be managed not only by UKBA (United Kingdom Border Agency) and its partner, G4S, but also by a well-known children's charity, Barnardo's. This development came despite promises from senior politicians that child detention had ended in the United Kingdom and despite international legal guidelines, such as The International Convention on the Rights of the Child, which stipulate that children 'should not be detained for reasons related to their migration status' (Farmer, 2013, p. 15). The new centre could be justified, however, because it was supposedly distinguished from ordinary detention via the degree of care it offered to residents and the 72-hour maximum time limit that children and families would be held for. 'This shift from security to care', write Tyler *et al.* (2014),

> was highlighted in a letter sent to local councillors and local residents by the planning consultancy company ... working on behalf of UKBA with regard to the CEDARS planning application. The letter stated: 'The facility will be run on a care model rather than a secure one, supported by a third sector organisation. In short it will look and feel very different to the UK Border Agency's immigration removal centres'
>
> Tyler *et al.*, 2014, p. 16

The centre organises occasional trips to the cinema and to the shops for families (under supervision), and visitors have reported that the centre provides comfortable beds, new furniture, spacious rooms and play areas. Indeed, the presence of the children's charity probably goes a long way to ensuring that the centre is operated carefully. Yet as Francis Webber, a practitioner of immigration law for over 30 years, confirmed, contrary to the government's claims that Cedars constituted a new, distinct class of pre-departure accommodation, Cedars 'will be a detention centre – complete with 2.5-metre perimeter fences, locked areas, internal fences dividing the site into accessible and inaccessible areas, a "buffer zone" inside the perimeter fence, and powers to use force and "control and restraint" techniques on both adults and children' (Webber, 2011).

Research has shown that it is detention itself that causes damage to children, rather than its conditions. The charity Medical Justice (2010) conducted research with 141 children detained between 2004 and 2010 in the United Kingdom for an average of 26 days each, and found that spells in detention were associated with increased risks of psychological harm (including developmental regression and expressed suicidal ideation) and physical health problems. Moreover, the manner of entry into, and exit from, detention outweighs the importance of the conditions within detention themselves. Of the 61 children who entered detention as a result of an unannounced dawn raid on their home, for example, over two-thirds were 'reported to have exhibited behavioral changes including panic, anxiety, and trauma' (Medical Justice, 2010, p. 5) afterwards, which often persisted after their release from detention (over 60% of the children in the study were eventually released back into the community). 'Play facilities and smiling faces', Webber concludes, 'cannot compensate for the lack of freedom' (Webber, 2011).

There are two salient characteristics of care in these extreme sorts of situations. First, care begins to act as a smokescreen for an immoral or brutal system. Typically this smokescreen will co-opt a set of ethical values in the work of a subjugating political structure. The first step down this road is to uphold the Kantian presumption of a rupture between 'the public virtue of justice and the private virtue of goodness' (Smith, 2000, p. 97). Once separated, the latter can be put to use in the service of the former as a way to licence inequitable or exploitative social and economic stratifications. Recognising this risk, theorists of care have warned that 'care needs to be connected to a theory of justice' (Tronto, 1993, p. 171). The mistake of carers who allow themselves to become part of brutal systems is to accept the primacy of the ethical over the political, of care over justice, as if any political machinery can be excused on the basis of an ethical or 'compassionate' approach.

The involvement of the children's charity Barnardo's in Cedars has been described as lending the UK Border Agency a 'cloak of legitimacy' (Webber, 2011). Other organisations held back from similar involvement that could lend credibility to some of the worst characteristics of the detention system. 'UK borders asked [my organisation] to provide training to their doctors in the recognition of torture', one medical doctor who visited detainees told me. If his organisation had agreed, then there is a high chance it would have lent legitimacy to the inadequate procedures for detecting torture survivors in detention discussed in the previous chapter (see The Independent Chief Inspector of Borders and Immigration and Her Majesty's Inspectorate of Prisons, 2012, for concerns over this inadequacy). 'However,' he continued,[26]

> [we were] concerned because we then went to an organisation which has done training of that kind internationally ... and they said 'Don't touch it with a barge pole. When we did training in [country] and [country] all that happened was the doctors rubber stamped "not tortured" on everything and justified their expertise in so doing on the grounds that they had received training from [us]'.

The second characteristic of care that has been co-opted within a subjugatory system is that the carers themselves may not view what they are doing as immoral: either they do not appreciate their own culpability, or they calculate that their co-option is a price worth paying – a necessary evil – in the protection of vulnerable people. Although laudable to a point, for Hannah Arendt it is precisely this sort of calculation that can ensure the cooperation of benevolent individuals in malevolent systems (see Arendt, 1964/1994).[27] For Weizman (2012) this same calculation of enduring a supposedly smaller evil in order to avoid a larger one has become a threatening generalised habit of thought among Western countries' military elite. The result is what Weizman calls 'humanitarian violence' according to which a measure of violence is understood as normal and necessary even among those concerned to minimize violence. As Weizman (2012, p. 3) puts it: 'the moderation of violence is part of the very logic of violence'. The role that 'well meaning citizens' (Weizman, 2012, p. 35) might play in collaborating with subjugating practices and systems is starkly foregrounded by Weizman's contemporary application of Arendt's work.

The result is the morally ambiguous *simultaneity* of compassion and repression (Žižek, 2009). The chief executive of Barnardo's offered compelling arguments when the charity took up its place in co-managing the new so-called pre-departure accommodation. In an interview with the *Guardian* newspaper in 2012, for example, she was able to point to a series of improvements in the way the government and G4S conduct their operations at Cedars because of their (Barnardo's) involvement, especially around the transportation of families, even though she was also quoted as saying 'Do I wish the [pre-departure accommodation] didn't need to exist? Absolutely' (Williams, 2012). One of her concerns was over what Cedars might become if the charity ever withdrew. 'If not us, then who?' she asks, implying that neither G4S nor the government have the ability to care as effectively as her charity for the families held there. In 2014 the charity pointed out the fact that, before Cedars, families and children were held for much longer periods of time, in worse conditions and with much higher frequency (Barnardo's, 2014). Barnardo's have also made it clear that, should practices at Cedars cross any 'red lines' that the charity insist constitute minimum humanitarian standards, they will withdraw from the running of the centre.

The gist of Barnardo's justification for its own involvement in Cedars is therefore something like 'without us, the treatment of suffering others would be even worse'. This type of argument, however, raises a thorny moral dilemma because, without the charity at Cedars planning permission may not have been granted for the centre in the first place (see Tyler *et al.*, 2014). What is more, should the charity withdraw from the running of Cedars then the government's reputation in the treatment of families facing deportation would be severely tarnished, potentially providing grounds for a wholesale re-evaluation of the system of detention and deportation of families, and yet no withdrawal has taken place.

Ultimately the involvement of kind, caring and compassionate people and organisations in subjugating systems of control is a fault of those very systems of control. It is very difficult, and probably sanctimonious, to ascribe blame or fault to caring actors under a brutal system. Nevertheless, the legitimacy they lend to such systems, and the degree to which they can be co-opted, is a profoundly insipid and tragic characteristic of dehumanising bureaucracy as well as one that the key theorists of moral distance and indifference risk overlooking by downplaying emotions.

Another way to understand the dynamics of the co-optation of care is to distinguish between empathy, compassion and care. Whereas it makes very little sense to include compassion in the list of values that places like Cedars attempts to embody, it makes more sense for them to claim to be empathetic. For Krznaric (2014, p. x), empathy involves 'the art of stepping into the shoes of another person [and] understanding their feelings and perspectives'. Empathy, then, primarily involves imaginative perspective-taking and does not, as such, entail a judgement that the situation giving rise to the feelings of the other person is undeserved. In comparison, for Nussbaum (2001) at least, for an individual to feel compassion they must believe that someone has suffered serious misfortune that they have not brought upon themselves, and that in the same situation this could have happened to them. Hence, 'empathy is not sufficient for compassion' (Nussbaum, 2001, p. 330).

With this in mind it is perfectly possible for UKBA to promote empathy whilst maintaining that a sub-subsistence income, detention or deportation of individuals is essentially something they deserve for having infringed immigration rules. One senior immigration manager explained her empathetic, but uncompassionate, position as follows:

> If I was a Nigerian woman, I would want to bring my family over here and live, I wouldn't have any right, so I'd have to be removed ... that Nigerian woman who comes over, she has to be removed, that's the policy ... but we can do that humanely and kindly and, you know, [ensure] goodness of service[28]

Although insisting on the correctness of the policy, the manager displays an understanding of the desire to migrate and the need for kind treatment. One spiritual support worker at Campsfield Removal Centre occupied a comparable position. During an interview with a local newspaper he reflected upon his difficult work (Paveley, 2005). 'Often I'll be walking around and will see someone who looks as if they are about to collapse in tears', he commented. 'I'll grab hold of them and just about get them into a room and the door shut behind us and they will be crying. At one point I lost so many handkerchiefs my wife was annoyed with me!' But despite this evident caring attitude that indicates a high degree of empathy, he made it very clear in the same interview that he felt that many detainees were justifiably detained and that they had brought their situations upon themselves:

> Many of the detainees are economic migrants ... Lots of [them] come in here and see me and say, you are a man of God, you should get me out of here and I say 'No, that isn't my job, that's your lawyer's job' ... you have to remember, lying is the lingua franca here, many of these detainees have paid huge amounts of money to get over here and are desperate to stay. (Paveley, 2005)

To be sure, compassion has its own drawbacks. The act of compassion can reproduce 'asymmetric relations', which can ultimately render it condescending and power laden (Korf, 2007, p. 370). The compassionate person often acts from a secure position, and can perform their privilege in the process of acting compassionately (Barnett, 2005). And where compassionate acts are trivial in the face of the suffering of others, they can take on a hypocritical character. 'Charity is the humanitarian mask hiding the face of economic exploitation', writes Žižek (2009, p. 19), whereas Fassin (2005,

p. 362) identifies the 'compassionate repression' that undergirds much of contemporary humanitarianism.

But the distinction between empathy and compassion remains important because compassion entails recognition that the suffering of another person is unfair: their suffering occurs through no fault of their own. There is, then, a sense of injustice inherent to compassion that empathy lacks, and that may provide the basis for a broader systemic critique of the social and political causes of injustice. For Nussbaum (2001), this relationship of compassion to wider notions of justice sets it apart from emotions such as pity, kindness, mercy and empathy, which are less likely to entail critical awareness of the political forces operating beyond the immediate encounter with suffering. It is for this reason that I return to a discussion of compassion in the context of activist struggles around British immigration controls in the next chapter.

When it comes to care, it is unclear whether the sorts of relationships the Home Office label as caring ones actually qualify as care. A range of definitions have been employed by different authors, from the person-to-person provision of services (England and Folbre, 1999, p. 40) to a 'sustained and/or intense personal attention' (Zelizer, 2005, p. 162) and 'an emotional bond, usually mutual, between the caregiver and cared-for, a bond in which the caregiver feels responsible for others' well-being and does mental, emotional, and physical work in the course of fulfilling that responsibility' (Hochschild, 1995, p. 333). Deciphering whether or not the institutions of British border control are 'really caring' therefore turns upon the particular definition of care at hand: the more expansive the definition the more poorly it appears to describe the sort of relationships that the Home Office refers to. The risk of a narrow definition, on the other hand, such as England and Folbre's, is that narrow definitions can reduce 'those cared for to physical objects on which work is performed' (Gill, 2012, p. 127).

Some prominent issues here include: (i) the extent to which care is tradable and marketable, because it has been suggested that it is precisely the act of going-beyond the tradable relationship that is the most valuable aspect of care (Goffman, 1961); (ii) the extent to which the care is mutual, because it has been suggested that genuine care affects the carer as well as the cared for (Hochschild, 1995; Bondi, 2008); and (iii) the degree to which care is needed as opposed to imposed, because, at its worst, care can sometimes be an exercise in the 'affirmation of relations of dependence and vulnerability' (Darling, 2011b, p. 412; see also Watson et al., 2004). On all three counts the view that the Home Office takes of care appears decidedly fragile. It is contracted, necessarily and intentionally non-mutual to the extent that officers are advised not to enter into relationships with detainees and other recipients, and imposed to the extent that the need for care itself has often only been created as a result of the enforcement of immigration controls in the first place.

Conclusion

In this chapter I have critically assessed the claim of a range of theorists of moral distance and indifference that indifference is essentially emotionless. I have argued that a variety of emotions are common in the everyday lives of immigration control functionaries in the United Kingdom who nevertheless remain indifferent to the suffering of many migrants that they manage. Although a minority of these emotions work against moral estrangement, many work to ensure it. This argument challenges received notions about the separation and opposition of emotions and bureaucracy that runs through the work of Simmel, Bauman, Weber and Glover, and constitutes another insight into the generation of moral estrangement in immigration control that I have developed in previous chapters.

The emotions of immigration staff are often conflicted. At times the immorality of immigration controls becomes patently clear to functionaries, and most ordinary people in these situations feel uncomfortable and ashamed at being part of such a system. At these moments, though, there are often powerful countervailing emotions such as anxiety that act to overrule or impede compassionate action. Furthermore, empathy, care and compassion are actively encouraged within the bureaucracy although it is unclear to what extent the latter two concepts accurately describe the sort of emotional responses among its functionaries that the Home Office labels as such. In such situations the system of immigration control is capable of co-opting these softer emotions, which presents an acute ethical dilemma to individuals and organisations within the system. At its most sophisticated, this co-optation of empathy, as well as the language of care and compassion, can be powerfully seductive in providing a justificatory story that functionaries and contracted agencies can tell themselves: they are acting sensitively and morally in the execution of difficult but necessary and unavoidable work. The subtext that this position endorses, unfortunately, is that the system of detention of children, of enforced poverty and of life-threatening deportation is here to stay, which constitutes a damaging vote of no confidence in the struggles of abolitionist groups.

The chapter has also provided grounds to prise apart some of Bauman's concepts relating to moral estrangement. Whereas Bauman generally elides moral insensitivity and moral indifference, this chapter provides at least two reasons to be cautious in doing so. First, given the conflicted character of emotions within the bureaucracy, according to which shame is overridden by other emotions like anxiety, indifference rather than insensitivity is a more appropriate metaphor. And second, where care, compassion and empathy are enrolled in exclusionary processes of border control, it is perfectly possible for functionaries to be highly sensitive towards the immediate plight of suffering others and yet remain indifferent in the sense that although they regret the suffering of others they see no reason for the political and legal systems that gave rise to this suffering to change. They either position the cause of the suffering outside their moral frame of reference or see it as morally justifiable, even as they express concern over the suffering itself. Their chosen course of action is therefore to be passionately impassive, caringly repressive or, to fuse Bauman's terms, sensitively indifferent. In a similar vein, Bauman and Donskis's (2013) book length discussion of 'moral blindness' does not do justice to the disturbing situation in which functionaries in immoral systems of control are anything but blind to the consequences of their work, recognise the immorality of the systems themselves, and yet determine that their actions within these systems are not only defensible but morally laudable as a 'price worth paying' or a 'necessary evil'.

Notes

1   Complete with an annual meeting – the *International and Interdisciplinary Conference on Emotional Geographies* – and a journal – *Emotion Space and Society*.
2   Interview with activist, 28 February 2006.
3   Conversation with immigration officer during the counter-protest, 27 July 2013.
4   Conversation with immigration officer during the counter-protest, 27 July 2013.
5   Interview with police officer, July 2006.
6   Interview with activist and community leader, 28 February 2006.
7   Interview with activist, 10 October 2006.
8   Interview with senior manager of the IND, Apollo House, 2006.
9   Interview with senior manager of the IND, Apollo House, 2006.
10   Interview with senior manager of the IND, Apollo House, 2006.
11   Interview with senior manager of the IND, Apollo House, 2006.
12   Interview with activist, 10 October 2006.
13   The comments referred to the judge's personal views on how easy it was for gay people to claim asylum in the United Kingdom.
14   Written communication with immigration judge, 21 July 2010.
15   There were actually two sex-for-visas scandals relating to workers in Lunar House that broke in 2006. In the first, which broke in January 2006, Anthony Pamnani, a former immigration officer based at Lunar House for four years, told *The Sun* newspaper that some immigration officers exchanged sex for immigration status. The claims prompted a formal inquiry but on 14 March Tony McNulty, then-Minister of Immigration, reported to Parliament that 'I am pleased to say that the investigation found no evidence to support *The Sun's* central allegation that there was a corruption "racket" in the Public Enquiry Office involving "sex for visas"' (Hansard, 14th March 2006). Although the inquiry revealed '[s]ome isolated incidents of unprofessional behaviour' (Hansard, 14th March 2006) and 'misconduct by some members of staff' (Hansard, 14th March 2006) it also emphasised the difficult job that staff at the IND did under extreme pressure. Later that same year, however, chief immigration officer James Dawute, also based at Lunar House, was apparently recorded offering immigration status for sex to a teenaged Zimbabwean female immigrant (Doward and Townsend, 2006). The interview with the senior manager

from which the quoted material in the text is taken refers to the first scandal and was conducted before the second.

16 Interview with senior manager of the IND, Apollo House, 2006.

17 Interview with attorney based in the United States, 19 August 2011.

18 Interview with immigration advocacy worker based in the United States, 30 June 2011.

19 Interview with attorney based in the United States, 30 June 2011.

20 Interview with charity worker based in the United Kingdom, 25 September 2011.

21 Interview with refugee advocate, London, 6 July 2011.

22 Interview with activist, London, October 2006.

23 A similar claim is made in Lampedusa, which hosts the 'temporary stay and assistance centre' (Andrijasevic, 2010, p. 148).

24 Interview with senior manager of the IND, Apollo House, 2006.

25 The advert reveals that there is a 'four week training course' for detention custody officers (G4S, 2013a).

26 Interview with doctor who had visited immigration detainees, October 2010.

27 Arendt was writing in the context of the Holocaust in Nazi Germany. While I avoid drawing direct parallels with the co-optation of care during the Holocaust, because I do not want to be misinterpreted as claiming that immigration detention resembles the Nazi death camps, there is a literature on the co-optation of care in Nazi euthanasia centres that illuminates a series of analogous situations and dilemmas to those I discuss in this section (see Benedict, 2003; Lagerwey, 2003).

28 Interview with senior manager of the IND, Apollo House, 2006.

Chapter Seven
Examining Compassion

In Chapter Four I described the work of South London Citizens (SLC),
a collection of schools, churches, mosques, synagogues, charities and residents' groups committed to taking action for the common good of the
people of South London, who investigated and wrote a report about the
conditions inside Lunar House. The following is an extract from an account
of the launch of their report into the conditions at Lunar House.

At our Assembly in Autumn Mrs. Lin Homer was our main guest. She is
now the head of the Immigration and Nationality Directorate [IND], and has
ultimate responsibility for Lunar House. Her presence at the assembly was the
fruit of a relationship which we have built up over the past few months. She
welcomed our report into the conditions at Lunar House and made a promise
to work alongside us for the common good. The Minister of State at the Home
Office, Tony McNulty, took and looked at our report. The Minister warmly
welcomed the report, acknowledged the thoroughness of it, and also the
nuisance we had been at times. This is a powerful symbol of just how far we
have come in the past year, and we thanked the Minister for his recognition of
us. Because of that recognition, Mrs. Apragas has a voice which can now be
heard. Two years ago she was ejected from Lunar House, shouted at and told
to go away. She was not believed. At the assembly Mrs. Apragas was able to
give a copy of the report to Mrs. Homer, as a sign of our appreciation that
Mrs. Homer has agreed to work with us, but also on behalf of the seemingly
powerless people who use Lunar House and feel ignored, misunderstood or
worthless and who this report champions, demanding that there is truly a
'humane service for global citizens'. In Mary's own words in broken English

Nothing Personal?: Geographies of Governing and Activism in the British Asylum System,
First Edition. Nick Gill.
© 2016 John Wiley & Sons, Ltd. Published 2016 by John Wiley & Sons, Ltd.

she said 'On behalf of South London Citizens I am proud to give you this report. Here is my story, but also the stories of many other people. It explains what our experience is like, and we hope so much that you will allow us to work with you in the year ahead so that other people don't have the same bad experiences. It means so much to have you here with us, thank you for coming and may God bless you'.

Email communication with South London Citizens worker, 2005

It had taken the best part of two years to bring the head of the IND face to face with Mary Apragas. The campaign had involved hundreds of people giving evidence, writing submissions, organising hearings and calling demonstrations. Then in 2009, four years after this Assembly at which SLC's report into the conditions at Lunar House was launched and handed to Lin Homer, a local newspaper reported that 'a new waiting area at Lunar House has been completed after years of discomfort for thousands of immigrants and asylum seekers who "waited in the cold and rain" to get seen by officials' (*Croydon Guardian*, 22 January 2009). This waiting area replaced the 'pig-pen' style railings and queuing system discussed in Chapter Four. 'The waiting facilities … have been upgraded by the UK Border Agency after the South London Citizens, an immigration campaign group, produced a report highlighting the need to improve conditions for the public', the newspaper article continued. Its construction was hailed as an '£800 000 victory' for South London Citizens (South London Citizens, 2009).

It is all too easy to dismiss the achievements of SLC at Lunar House. Lunar House is still the site at which terrifying and subjugating border practices are meted out against vulnerable migrant populations despite a decade of activism at the site. Although the SLC team managed to secure the cooperation of senior figures in the IND to implement the recommendations of their report in 2005, the senior figures themselves were often truculent. What is more, the list of demands SLC made could be seen as parochial and modest, calling for better waiting facilities rather than a radical overhaul of immigration controls. For some of the activists involved in producing the report, however, the material victories, like the covered waiting area, were less significant than the symbolic victory won at that autumn Assembly back in 2004. 'I sometimes think that there will be someone down the road who will write a PhD thesis or write a publication saying that we really sold out when [we] did that report because its liberal rubbish and it doesn't really change anything', one of the activists mused shortly after the Assembly, eyeing me suspiciously.[1] 'Well maybe we did,' he continued,

but I tell you what, when Mary [Apragas], the woman which it all started from who had her papers lost in Lunar House, … goes up to Lin Homer and said 'Two years ago you took – the system took – my voice away', and she handed the report to this most senior civil servant of the Home Office and said 'thank you for giving it back'. And she gave her the report. Now that may be just good political theatre but I just thought, actually, regardless of what's happened, regardless of what happens next, in and of itself it was worth it. Just for that one moment.

It is this purposeful overcoming of the artificial keeping-apart of decision makers and migrants that this chapter examines and in particular the pursuit of more compassionate officials giving rise, in theory, to fairer and more humane immigration controls. In the previous chapter I established that feelings of compassion entail a judgement that the misfortune that has befallen an individual is undeserved (Nussbaum, 2001). As such, compassion embodies a feeling that injustice has occurred, which is not necessarily the case for empathy, nor pity, and which might imply that the pursuit of compassion among functionaries is a worthwhile objective. This chapter consequently critically assesses the work of migrant support groups who seek to nurture compassion among functionaries and managers.

For the SLC activists, the detachment of managers from the human consequences of the work that they oversee highlights, first and foremost, the difficulty and novelty of bringing elite system managers like Mrs Homer face to face with one of its users in a situated encounter (as noted in Chapter Two, this is even less likely now than it was in Mr Brimelow's day). The achievement of these activists, quite apart from the material upshot of their struggles, was to bring senior civil servants and government ministers into proximity with the marginalised and excluded, into a situation of 'being addressed by the appeal or accusation of the Other' (Barnett, 2005, p. 19). This was understood as a way to nurture deeper understanding among elite bureaucrats in the hope that they will be moved to act compassionately and humanise the border controls that they manage.

In what follows, however, I critically examine compassion. I begin by showing how popular the pursuit of compassionate functionaries and system managers through nurturing fuller understanding of migrants' situations has become among asylum support groups in the United Kingdom. I demonstrate that the logic of compassion nurturing is identifiable at a variety of scales ranging from activists' media work to small-scale, one-off activist interventions within detention centres. Following this I discuss how distinctive the pursuit of compassion is by comparing the seeking of compassion to other forms of non-revolutionary activism. It transpires that activism in pursuit of compassion is distinctive because it is often carried out on behalf of migrants (rather than by them), it is overt in the sense that it is carried out in full view of the existing authorities (rather than surreptitiously), and it tends to accept the existing configuration of powers that bureaucrats are at liberty to exercise (precisely in order to entreat them to exercise their power compassionately). In this light, the case for scepticism about the potential of compassion to be truly emancipatory becomes clear. Activism in pursuit of compassionate functionaries often entails a degree of capitulation to existing configurations of power that may be unpalatable to activist groups, even including those pursuing non-revolutionary, non-abolitionist forms of activism.

In the face of these observations, defending compassion is difficult. Although my interviewees do recount some instances in which compassion has functioned as an effective way to curb the worst excesses of the bureaucratic control of borders, especially in cases where local efforts have 'scaled-up' to system-level initiatives (and I return to the SLC case in this context), these accounts are not enough to overturn the damning charge of the capitulation of compassion to existing systems and configurations of control. At best, they may be enough to warrant a milder view of compassion in specific circumstances when alternatives have been exhausted. In other words, although it is possible to list a set of mitigating considerations that can allow compassion to be an effective activist response in certain circumstances, all in all the case against compassion is powerfully compelling and indicates that compassion seeking as an activist tactic must be approached extremely cautiously. I outline a catalogue of risks that an approach that seeks explicitly to nurture compassion runs, I argue that the appeal to compassion is a treacherous route to improving the lives of asylum seekers in the United Kingdom, and I conclude that it should be seen as the activism of last resort as a result. Although the recent evolution of border controls can be seen to have squeezed the humanity out of border controls, it transpires that seeking to (re)instate humane borders is not the best way to counter this development.

Closing Moral Distance Through the Media

In this section and the next I establish how widespread compassion-seeking as an activist tactic has become, first at the general scale of 'the public', then, in the next section, through the lenses of interaction and sustained contact. A primary example of work to nurture compassion is media-based activism. Media-based responses to moral distancing target 'the public' in general as much as they do immigration employees and functionaries, but their purpose is to furnish those who are unfamiliar, poorly acquainted or misled about refugees and asylum seekers with rich and realistic accounts of refugees' experiences. Consequently, this sort of work tends to emphasise the human content of the media-based narratives that it produces and disseminates. 'I just think the impact of having refugee voices in the press speaking from their own personal point of view has an impact on not just journalists but public attitudes as well', one management-level charity employee told me.[2] Often these stories are seen as very different in character to political lobbying or policy approaches, which are perceived as lacking personal specificity. 'Our organisation is focused on working through social media and news media, raising awareness of detainees and by extension giving people a chance to think about their attitudes towards migrants in general but through a human interest sort of approach rather than through a policy approach', the manager explained.[3] 'Political lobbying tends to appeal to people who are already interested in political lobbying', another activist observed,[4] 'but a lot of the general public are probably turned off by political lobbying and you need the human interest thing to really draw people's attention'.

Emphasising the personal through individual stories inevitably raises concerns over revealing important details of individual migrants' lives and exposing them to unwelcome attention from the media or border control enforcement authorities. Some media-focused groups, however, demonstrate a high degree of confidence that they can use personal stories safely. 'We anonymise cases', I was told by one interviewee who worked as a communications officer for a large refugee support group,[5] 'and I think we're quite good at removing sufficient detail to make the case and the person completely unidentifiable'. Indeed, it is the 'human interest angle' of the story that is seen as the best antidote to dehumanised and hostile portrayals of asylum seekers in the press. Working class communities, for example, are understood to be more likely to respond positively to stories about migrants when they can relate migrants' stories to certain aspects of their own lives. 'You'd do more for the cause of people liking migrants if you can establish them in the committee of members', one activist explained,[6]

> like they both support Manchester City or something. It's all about seeing common humanity. So stories can be woven into that. Someone in a sink housing estate can relate to the idea that someone might want to move away from where they've grown up in order to better their lives and find better for their families. They can actually relate to that really well.

Furthermore, whereas it is generally accepted that there is not much mileage in appealing to confirmed right-wing nationalists through the media, the human interest in stories is seen as the most effective way to influence those who are still unsure about migration-related issues. 'So your *Guardian* reader, well we don't have to worry about them because they're onside', one manager of a refugee support group that worked to raise awareness about immigration detention in the United Kingdom told me.[7]

> And the *Daily Mail*, well forget it, you're not going to change that overnight. But there's that undecided middle who have ideas but not firmly established opinions about migrants and migration and asylum and detention and all that stuff. That's where the human interest thing can maybe have an impact.

The desire to directly confront and repair negative media stereotypes and prejudice is also palpable among asylum seekers themselves, who often find the label 'asylum seeker' degrading because of the way the right-wing press has represented them. 'I don't want to be living as an asylum seeker because I have to be showing myself as an asylum seeker', I was told by one Kurdish man who had been waiting for a decision on his claim for three years;[8]

'I'm so embarrassed because of what has happened now in the media. Since I came to this country we have been always attacked by the media, many newspapers in this country, they are against morality, against humanity. They are against us.' Including personal details of the reasons for migration and migrants' struggles against poverty in the United Kingdom is seen by many activists as a way to encourage journalists and newspapers to employ more balanced reporting and is commonly cited as a way to reduce prejudice.

Concern over ignorance about refugee and asylum seeker issues is especially acute in the case of young people, with some groups seeing work with school children specifically as a key investment in a more compassionate society in the future. Nussbaum (2001) discusses the importance and feasibility of an educational 'imaginary curriculum' (Nussbaum, 2001, p. 433) that exposes children to accounts of undeserved misfortune as a way to construct a more compassionate society. 'The younger ones they have the jobs tomorrow', I was told by one refugee-led asylum support group volunteer who spent a large part of his time visiting local schools.[9] 'If we start creating that awareness among the young ones, it will not have an effect now but in the future'. 'There's a presentation that we adapt for schools', another asylum support group worker[10] told me.

> Last week I asked what percentage of the population do you think are refugees and one person said 25 per cent! I've done a few at the Sixth form college, just like, 'You know what it feels like to be a refugee and asylum seeker?' and things like that. We also cover the difference between refugee and asylum seeker and the barriers they might face, not just language but legal barriers or emotion or trauma or that kind of thing.

Although a lot of the media work that asylum support groups are engaged in targets broad publics such as 'young people' or the 'undecided middle', rather than presently employed asylum sector decision makers, my interviewees were not short of theories about how changing public opinion or targeting mass communication channels could improve the ways in which government officials specifically interact with migrants. They pointed out that officials are not impervious to the newspapers and reflected that broadcasting may be their best chance of communicating with some of the most insulated officials. For some of the larger refugee support groups, the importance of positioning themselves so that they were in conversation with government through the press was also important. '[We've] been very successful in terms of press work', one refugee support group manager explained,[11] 'in that in the past you would get the press cuttings for the day and the Home Office would issue a press release and they would have been copied verbatim in ten different newspapers. Whereas now and particularly in the Scottish press they will always approach us for our view or to seek the view of a refugee or for a case study'. When press releases from the Home Office appear alongside alternative, more humanised, depictions of individual stories Home Office employees are offered the opportunity to interpret the story independently and begin to think more critically about their roles.

Nurturing Positive Interaction

Other groups employ a more targeted approach that is less focused upon awareness raising among broad publics and more focused on establishing positive interactions between decision makers and their subjects. This work directly confronts the mechanisms of psychological distancing at close quarters identified in Chapter Four by establishing and nurturing meaningful, respectful personal interaction between decision makers and subjects. In situations where interaction is rare, or else stilted, fleeting, unnatural or constrained, such as during asylum interviews, various migrant support groups in the United Kingdom work to humanise and personalise the immigration system by nurturing fuller interaction between border officials and migrants. In other words they foster familiarity, which recognises that stories, accounts and self-expressed narratives of migrants need the 'space to breathe' (Woolley, 2014). This sort of approach is based on the view that asylum seekers' narratives cannot be straightjacketed into rigid, formulaic mediums of exchange and still retain their impact, their power or even their full meaning. 'In telling and listening, narratives create meaning and help to make sense of our lives', writes Smith (2015, p. 463). The objective of these sorts of tactics is consequently to loosen the strictures that confine the narratives that migrants are able to share. The aim is to ensure that they are not merely recognised as part of this or that group or category, but that their experiences and sufferings are acknowledged on their own terms. For the SLC activists, for example, it was about Mary Apragas's *voice* – her ability to relate her grievances and her frustration in her own way.

Whereas the SLC campaign was focused upon a specific set of issues around Lunar House, Living Ghosts was a national campaign that ran during the mid-2000s with the aim of giving destitute asylum seekers the opportunity to tell their stories to influential figures in their communities who rarely come into contact with asylum seekers. The campaign was organised by Church Action on Poverty, a national ecumenical Christian social justice charity, and invited activists to take part in various forms of solidarity with destitute asylum seekers, including the 'asylum destitution endurance challenge', which involved living off £5 and a food parcel for a week, as well as through prayer, writing to local MPs and donating funds. For those who wanted to 'go further', the campaign pack also provided detailed instructions on how to stage a 'poverty hearing':

> The concept of a Poverty Hearing is as simple as it is challenging. The main purpose of a Poverty Hearing is to provide the opportunity for people with

direct experience of poverty to speak out for themselves while encouraging those with power, authority and different experiences to listen. Much of the local and national government plans to alleviate or eradicate poverty is done by people who know about deprivation by reading or hearing about it, but not necessarily feeling it. Part of the reason for injustices is that people with direct experience are deliberately excluded or not meaningfully involved in developing remedies. For poverty really to be tackled it has to be built on the expertise of those who know about it first-hand.

Living Ghosts Action Pack, 2005

The action pack emphasised that 'those in poverty are the real experts' and recommended that invitations be sent to MPs, church leaders, counsellors, business people, trade union representatives and other public figures to attend regional hearings that were held around the United Kingdom in the mid-2000s. Destitute asylum seekers featured heavily among those in poverty. The aims of the campaign, the action pack explained, were to 'resource sympathetic people' who want to change the political climate and policies that cause destitution as well as to 'press government to allow refused people seeking asylum to work and/or claim benefits'.

Determining the success of the campaign is very difficult. On the one hand, the organisers can point to some important policy changes that occurred during the mid-2000s, including abandonment of the blanket outlawing of paid work for asylum seekers following legislation from the European Court of Human Rights in 2005,[12] and the fact that asylum seekers also now, technically, have access to welfare benefits even if they have been refused.[13] On the other hand, although these developments might be hailed as victories the stipulations, conditions and limitations imposed on both the right to work and emergency support for refused asylum seekers lessen their significance in practice.

At a local level as well, the success of the campaign is unclear. The poverty hearing in the South-West, for example, which was held in Bristol in early 2006, was hailed by its organisers as a modest success, but not quite what they had hoped for. Although it had good attendance, many of the key people they had wanted to target did not arrive. One local television company had said they would be there but did not attend, and a local newspaper covered the event the week before but not on the day. Another key journalist and many of the MPs that were invited also did not attend, largely because, as one of the organisers ironically reflected, they saw it as 'too political'.[14]

Nevertheless, the approach Living Ghosts took in order to exert pressure over the government embodies a distinctive and innovative methodology that emphasised staged interactions between marginalised and influential individuals that are often either impossible or very rare under the normal conditions of bureaucratic administration. Work that short-circuits the way decision makers and subjects are kept apart directly challenges the mechanisms of moral distancing that are inherent to bureaucratic forms of administration and, as an approach to activism, it is therefore significant due to the directness with which it addresses the impersonalism and moral estrangement described earlier in the book.

Other tactics target the interaction between decision makers and their subjects in different ways. Although not working to necessarily *establish* interaction like SLC and Living Ghosts, some groups are more concerned with improving the qualitative nature of the interaction in situations where decision makers may already be in frequent contact with migrants. Sometimes this involves giving close attention to the technical procedural elements of asylum interviews. For at least a decade, for example, the United Nations High Commission for Refugees (UNHCR) has chipped away at the procedural features that make an asylum interview, such as those conducted at Lunar House in the mid-2000s, 'a rigid and unbending context quite unsuited to capturing the complexity of personal experience' (Woolley, 2014) through both their Quality Initiative and Quality Integration programmes. These programmes include an agreement with the Home Office to allow the UNHCR access to observe asylum determination procedures. Through a series of reports, the UNHCR has made numerous recommendations about how interviews should be conducted and interviewers trained. These include recommendations for accreditation of frontline employees involved in making decisions, using a decision-template in order to ensure that key considerations are not overlooked by decision makers, requiring that all trainers of interviewers have recent case-work experience, observing interviews frequently, especially for those new to the job, and encouraging interviewers to focus their questions to avoid unnecessarily lengthy interviews (UNHCR, 2005, 2006b, 2007).

Various other recommendations concern the particular mechanics of interaction during the interviews themselves. Concerns have been raised, for example, over the lack of eye contact, the lack of a neutral and professional tone employed by the interviewer, a lack of opportunity given to interviewees to expand upon their stories and explain apparent inconsistencies, and the use of 'rapid-fire' closed questions (UNHCR, 2006a, p. 14; see also UNHCR, 2006b). Other concerns centre around the use of interpretation. 'In some cases the applicant was addressed in the third person', for example, 'with the interviewer saying to the interpreter: "Ask her if..."' (UNHCR, 2006a, p. 14). In the case of interviews with children, inappropriate seating arrangements have been identified as well as the lack of introductory material for the interview that was likely to be comprehensible to a child (UNHCR, 2009). Activities to address these concerns aim to nurture positive, or at least less negative and stilted, forms of interaction during the interview.

Whereas UNHCR's efforts centre upon the routine format of asylum interviews, other migrant support groups organise meetings between migrants and decision makers in different contexts to those in which they usually meet. One group, for example, explained that they 'have been working locally with

[United Kingdom Border Agency (UKBA) officials]. We've been very pleased actually that we've been able to go and talk to them. They invited us in to do some training. It was a couple of refugees, who are working in community organisations now, just talking about how they are putting back into the community and how they are integrating and stuff like that'.[15] By changing the content and purpose of the discussion – away from interrogating the credibility of the claimant towards the contribution of refugees to the local host community – the intention of this group is to alter the tenor and rhythm of the interactions between decision makers and subjects.

Interestingly, the demand for these sorts of training events often comes from the decision makers themselves. Another support group that specialises in the care of victims of torture and was 'invited to participate in an away day for [Lunar House] staff' similarly reported the desire among staff to gain 'an insight into the customers that [Lunar House] serve and their experiences prior to applying for asylum in order to create an understanding of the vulnerabilities of the customer as a direct result of the trauma experienced in their home country'.[16] The same desire to encounter migrants differently was evident among the employees I interviewed. 'Even if we had like a course to show us what had happened in their country or where they were coming from that would have helped', the former intelligence officer based at Lunar House noted,[17] 'because I know that I didn't know a lot about why people were here. I didn't really know what was going on in [country] at that time or why there was, like, an influx of them. It would have been useful to educate us on where they had come from and their struggle.'

Again, it is difficult to measure the success of these initiatives. Certainly some elements of the interview procedure have been changed as a result of the UNHCR's efforts, for example, but equally the procedures remain traumatic not only because some of the recommendations UNHCR have made have been overlooked[18] but also because of the inherently intrusive nature of the asylum interview process.

A third genre of attempts to nurture compassion concerns the relationships between Detention Custody Officers (DCOs) and detainees in the immigration detention estate and addresses the overfamiliarity with suffering that I described in Chapter Five. In situations where compassion fatigue and secondary trauma among decision makers are constant threats, tactics aim to ensure that interaction between decision makers and migrants is not trained exclusively upon the trauma of an individual. The emphasis is again upon changing the dynamics, tenor and quality of their interactions, but, in detention, such tactics work to stage informal interaction between functionaries and subjects as a way to preclude or discourage overemphasising trauma. One detainee support group, for example, organises artistic sessions for detainees that focus primarily upon giving the detainees the opportunity for self-expression. But, as the manager of the group explained, 'There have been times when we have made a real difference to how some staff have interacted with the detainees as well. We've kind of enabled staff to think afresh about what they are doing a bit.'[19] By way of illustration he continued:

> There is an indirect effect of our work. [One of the managers] said that she felt she'd seen staff interact differently with detainees, even staff who hadn't been in the session with us, because they knew [the sessions] were happening and it gave them something to talk to the detainees about. Because they're familiar with a member of staff in the room during the [creating], that can change the relationship they have with the detainees. Because they've met on kind of an equal footing in the context of the [art]. [Some] staff want to engage with detainees but are not sure what to talk about; I mean what do you talk about? [During the sessions] there's something there that is positive and they can say 'Oh, did you do the [art]? Oh, great.' That sort of simple thing can have a ripple effect within the institution. So I wouldn't want to overstate it because there are many other influences on how staff and detainees behave and many of them are not at all positive. But to have some impact at least, to make some things better some of the time, even if that is all it was, it's good.

As I have shown in the previous chapters, the effect of the bureau-cratic administration of border control is to squeeze the middle ground bet-ween under- and over-familiarity with suffering others almost to vanishing point. Bureaucracies propagate two opposite yet complementary forms of indifference – one that estranges functionaries from trauma and one that saturates them with it. The delicate art of nurturing compassion is to prise open the intermediate territory. In situations of under-familiarity with refugees' stories, media exposure and meaningful contact might be nurtured. In situations of overfamiliarity with suffering, tactics that nurture compassion seek to ensure that interaction moves beyond the fact of difference and avoids fetishising suffering or experiences of disaster. This type of interac-tion acknowledges the subject as more than a victim and fuller than the misfortunes that have befallen them.

The Distinctiveness of Seeking Compassion

In the foregoing sections of this chapter I have demonstrated that the logic of nurturing compassion is widespread at a variety of scales. The common denominator of all these activities is that they strive, successfully or otherwise, to disrupt the mechanisms that lead decision makers to view their subjects as units, specimens and dehumanised elements rather than real people. Media work seeks to elicit acknowledgement and improve understanding of refugee experiences, whereas up-close tactics work either to establish interaction or to promote different, fresh ways to interact in order to improve relationships and humanise the business of rule. These types of activism are by no means unproblematic and in the next section I set out the case for scepticism about compassion-seeking and compassion-nurturing. Before I do, however, it is worth considering how distinctive these tactics are. In this section I consider where, if at all, activist efforts to nurture compassion fit alongside some of the more commonly discussed types of activism in the social sciences. In doing so, a fuller picture of the character of attempts to nurture compassion comes into view.

The attempts to nurture compassion that I have described are not generally revolutionary. Their objectives fall short of a complete system overhaul such as the wholesale disbanding of immigration controls, and so, on the whole, they are likely to disappoint radical left-wing thinkers who maintain that 'economic and political power ... will not be whittled away by a slow process of erosion, nor destroyed by a succession of partial reforms' (Gorz, 1968, p. 112). This is an important point. If immigration control is inherently unreformable owing to the excess of harm that it causes over any modifications to its organisation that can be made, then we may be forced to acknowledge not only that efforts to humanise and soften the system are fated to fail, but also that they offer an opportunity to legitimise a fundamentally exploitative and subjugatory arrangement.

Various thinkers, however, have defended some form of activism that works towards goals short of revolution. De Certeau (1984, p. 27), for example, dismisses the possibility that the subordinated classes are likely to rise up against the dominating system any time soon as 'fabulous, as in the stories of miracles'. However, he writes, although '[t]his hope has disappeared ... a diversionary practice remains possible' (de Certeau, 1984, p. 27). Lefebvre concurs when he writes that, while there is 'no doubt' that reformism was wrong (Lefebvre, 2009, p. 140), 'reformism has not been completely wrong. If it made no sense it would have disappeared. Its permanence cannot be baseless. An absolute rupture, a leap from necessity into freedom, a total revolution, and a simultaneous end to all human alienation, this doubtless naïve image can no longer be maintained' (Lefebvre, 2009, p. 140).

It remains less clear, however, where among the conceptual approaches to activism-short-of-revolution activities to nurture compassion should be located. Influential work has emphasised the importance of not simply lumping all non-revolutionary activist work together as 'reformist' (Apple, 1995, p. 120): there are different ways to undertake diversionary practice. James Scott (1987), for example, describes the struggles of slaves against colonial rulers, in which 'open defiance was normally foolhardy' (Scott, 1987, p. 33) but 'everyday forms of resistance' (Scott, 1987, p. 36) were both more common and more effective than 'the few heroic and brief armed uprisings about which so much has been written' (Scott, 1987, p. 34). These forms of resistance include 'footdragging, dissimulation, false compliance, pilfering, feigned ignorance, slander, arson, sabotage, and so forth' (Scott, 1987, p. 29) and constitute some of the 'weapons of the weak' through which indigenous communities could make the lives of their Western colonisers very difficult indeed. Yet these forms of resistance are carried out by the subjugated themselves.

The work to nurture compassion that I have described is largely undertaken on behalf of migrants rather than by them. Although former asylum seekers and detainees may be involved in these efforts and occasionally organise them, those still within detention or awaiting a Home Office or judicial decision are usually not in a position to take on leadership roles. The SLC initiative was led by prominent religious, legal and academic figures, for example, and although destitute asylum seekers gave evidence at the Living Ghosts hearings, they did not initiate them. Hence, although Scott's everyday resistance is non-revolutionary, its tactics differ markedly from attempts to nurture compassion in terms of the degree to which intermediaries are involved in acting for and speaking on behalf of the subjugated.

For de Certeau, the diversionary practice of which he writes consists of 'tactics', which he characterises as guileful 'ruses' (de Certeau, 1984, p. xv). The art of tactics entails '"putting one over" on the established order on its home ground' (de Certeau, 1984, pp. 25–6). Hence de Certeau's tactics involve situations in which the dominating system 'is tricked' (de Certeau, 1984, p. 26). The work to nurture compassion that I have described, however, operates through awareness raising and improving understanding of migrants among bureaucrats rather than through dissipation, evasion and trickery. As such, we should draw a distinction between de Certeau's tactics and work to nurture compassion because the latter is overt and upfront, rather than surreptitious, in its relationship with the dominating power.

Feminist thinkers have also wrestled with how to conceptualise non-revolutionary activism, partly in response to the machismo that besets notions of wholesale revolution, giving rise to a need to understand post-heroic forms of activism more clearly (Larner and Craig, 2005; see also Chatterton and Pickerill, 2010). In discussing the challenges that globalisation poses to children in poverty, for example, Cindi Katz distinguishes resistance, resilience and reworking as 'three creative strategies that people [use] to stay afloat and reformulate the conditions and possibilities of their everyday lives' (Katz, 2004, p. x), which acts as a 'useful counterpoint to the more typical labelling of all nonconforming action as resistance' (Holt, 2007, p. 1269). Resistance, the first type of response that Katz identifies, is fundamentally oppositional. It is the rarest of the three types because it is often more difficult and more costly than the others but it is the closest of the three to radical revolutionary activism. Resilience, on the other hand, describes those 'small acts' (Katz, 2004, p. 244) of 'getting by, finding new and creative ways of surviving' (Cumbers et al., 2010, p. 60).

In the middle of what can be seen as a continuum of reactions to subjugation (Sparke, 2008), reworking serves as a 'transitional category' (Cumbers et al., 2010, p. 61). Reworking is understood to 'alter the organization but not the polarization of power relations' (Sparke, 2008, p. 2).

Projects of reworking tend to be driven by explicit recognition of problematic considerations and to offer focused, often pragmatic, responses to them. They … are enfolded into hegemonic social relations because rather than attempt to undo these relations or call them into question they attempt to recalibrate power relations and/or redistribute resources. This is not to say that those engaged in the politics of reworking … support … dominant social groups, but that in undertaking such politics, their interests are not so much in challenging hegemonic power as in attempting to undermine its inequities on the very grounds on which they are cast.

Cindi Katz, 2004, p. 247

Katz's concept of reworking captures one of the key aspects of work to nurture compassion: the critical approach to, and engagement with, centres of power and systems of control on their own terms, rather than outright refutation (as in resistance) or evasion – as in Katz's (2004) resilience or de Certeau's (1984) 'tactics'. Reworking, then, like attempts to nurture compassion, is neither revolutionary nor surreptitious. Yet attempts to nurture compassion are, at best, a very specific type of reworking. They do not, for instance, attempt to 'recalibrate power relations'. The influence and position of the powerful is not explicitly put into question by the sort of activism by Living Ghosts, SLC or the UNHCR I have described. Indeed, it is precisely because the individuals that these campaigns target are so powerful that they became objects of activism: the powerful are appealed to, almost pleaded with, rather than challenged, and the existing calibration of power relations is taken as given. Furthermore, although there are minor skirmishes around resources, such as the SLC's attempts to secure a new waiting area, the real battles are around acknowledgement and understanding. The sympathies of decision makers, rather than the distribution of resources, are the real prize of efforts to nurture compassion. This sort of activism, then, targets the way decision makers think, their allegiances and dispositions, more than either the power or the resources that they command.

Neither Scott's everyday resistance, de Certeau's tactics, nor Katz's reworking therefore fully capture the key characteristics of efforts to nurture compassion, even though all four activities describe different ways to carry out activism short of revolution. Indeed, arguably we learn most about attempts to nurture compassion when these attempts are contrasted to, rather than equated with, existing conceptual approaches to non-revolutionary activism and resistance, as I have in this section. In so doing it becomes clear that attempts to nurture compassion are mostly mediated (i.e. undertaken on behalf of others), overt and sympathy-focused. In short, the attempts to nurture compassion that I have identified as operating throughout the mid-2000s in the United Kingdom around border control are distinctive, and as such they require an independent assessment.

Against Compassion

Various difficulties can beset attempts to promote conscientious and compassionate attitudes towards suffering others. Representing asylum seekers in the press, for example, not only introduces the risk of exposure of asylum seekers to unwanted attention, but also risks, as one communications manager of a small detainee visiting charity put it, 'hitting the general public over the head with the story of somebody else's trauma'. People go "Yeah, OK"',[20] he continued,

> but feel they're being told 'You've got to feel sorry for this person', 'You've got to be nice to them', because they've had this awful time. All they ever see of a detainee or a migrant is a victim, someone who doesn't have power, is not a real person, just this other. You don't want to hit them over the head with it because then they feel that they're being told what to think and they back off. And you sense the more outlandish the trauma that this person has gone through the harder it is for somebody to relate to.

Sometimes then, getting the 'personal angle' into the news story has unintended negative consequences like depicting the refugee as disempowered and passive (see Malkki, 1996, for a general discussion of this effect among work that attempts to advocate for refugees). Moreover, activist media officers and journalists are often tempted to find the 'most deserving' case study. As one communications officer put it,[21] 'My colleagues just want to work with a nice asylum seeker, somebody who has suffered properly, you know?' This creates a related risk around working with the press. The difficulties of representing and attempting to speak for marginalised others, even within discourse that is supposed to be radical and emancipatory, has long been recognised (Spivak, 1988). Dividing the asylum-seeking population into those who are deserving and undeserving shores up the legitimacy of the border control system to arbitrate claims because it applies a similar logic: there are some desirable migrants and some undesirable ones, and it is 'our' prerogative to choose between them. This can lead groups working with the media to skim over difficult issues in an attempt to make the 'featured' asylum seekers more palatable and media-compatible. 'Some of the detained family cases we are involved in have involved women engaged in an awful lot of criminal activity',[22] one support group worker outlined,[23] 'but obviously we're not drawing attention to that. So, you just have to pick your way through very carefully.' The same interviewee described how their work with foreign national ex-offenders, 'is not really for public consumption because there's a sort of double whammy. They are criminals and they are also foreigners and often asylum seekers as well, which makes it a triple whammy, so it's probably not really worth the effort trying to do proactive media work in that area'. The result is that some groups end up projecting a sanitised, partial view of asylum seekers accompanied by an often unarticulated acceptance that the less desirable or vulnerable asylum seekers can be legitimately excluded.

This issue comes to a head in relation to the issue of the pitiful refugee. There is a temptation for those working with the media, as well as legal actors keen to secure immigration status for applicants, to emphasise the vulnerability of certain refugees, especially women, children and sick refugees, because this has the advantage of eliciting pity for them. Pity is a close cousin of compassion but less connected to a theory of justice and injustice and more reliant upon the perceived wretchedness of the individual concerned. Although it may be tempting to project the pitiful refugee in certain cases, the overall effect of this tactic is to perpetuate the image of a passive, helpless, non-agential migrant figure, which often could not be further from the truth especially with regard to women escaping violence, migrants coping with ill health and unaccompanied minors making their own way to safety. Pity is also closely related to charity in its most condescending sense. It proceeds from graciousness and generosity (rather than concern and solidarity), which are dangerously optional on the one hand and rather flattering to the immigration control system on the other.

When it comes to nurturing positive interaction between decision makers and asylum seekers other risks present themselves. In certain circumstances staging contact between a decision maker and a subject can entrench rather than soften sceptical and standoffish attitudes towards migrants. As discussed in Chapter Four, not all contact is positive, even if it is organised with the intention of creating a morally obligating encounter or bringing decision makers into moral proximity with migrants. Interaction is 'messy' (Askins and Pain, 2011, p. 813) and whether worthwhile social change occurs as a result depends upon the form of the interaction itself (Allport, 1954).

Working to improve and soften the business of rule can also lead to incorporation into the very system that is being challenged. The work of the group providing art sessions to give detainees an opportunity for self-expression, for instance, may have the indirect effect of changing the dynamics of the interactions between staff and detainees, but also comes perilously close to providing 'an apology for the existing reality' (Lefebvre, 2009, p. 38). As the manager of the group explained:[24]

> There are frankly myriad injustices in the detention system and people feel very strongly about this, rightly so, and they want to change them so the logical step is to campaign on those issues and to try and raise them with government or with UKBA. And so we're a bit unusual, we're a bit different from all of that, because we ... work much more within a regime of detention than I guess most others. Which isn't to say that [we] don't go into detention but we're more part of the centre's own programme of activities and nestled in with that. So we, it's not so much on principle that we don't get involved in

lobbying but it's sort of a pragmatic decision not to get involved with lobbying. Because if we got involved with lobbying the level of access which we have into detention would be unlikely to be sustained.

This close cooperation with the management of centres opens the group to the charge of co-optation. As the manager continued:

> The detention centre gets quite a lot out of [the art sessions for detainees]. They get brownie points in inspection, they get less challenging behaviour from detainees … there are organisational drivers for them so, you know, they can justify the spend on it being good for their organisation and for their track record and things like that. Less riots.

The degree of closeness to the bureaucracy is a calculated and often agonising decision. Some artists even disassociated from the group, the manager told me, due to feeling differently about this very calculus. 'One [set of artists] got very closely engaged through [art] projects with detainees … and were interested initially in progressing that involvement', the manager told me, but 'then decided not to because they felt that to get more involved would somehow condone the work of the detention centre'.

What the experience of this group illustrates is the very real risk that attempting to humanise parts of the system or make them more compassionate might legitimate and bolster it overall. For Nussbaum (2001), depending upon compassion is both inadvisable and should not be necessary in a just society. She observes that the mechanisms that bring people to feel compassion and act compassionately are unreliable because they rest upon 'the senses and the imagination in a way that in principle makes them narrow and uneven' (Nussbaum, 2001, p. 386) and warns that societies that place a great deal of emphasis upon compassion as an organising principle can expect 'uneven and at times arbitrary results' (Nussbaum, 2001, p. 386). Compassion, she argues, is therefore 'an insufficient, and even a dangerous, moral and social motive' (Nussbaum, 2001, p. 361) and consequently that 'it would not be good to rely on it too much' (Nussbaum, 2001, p. 387).

It is a measure, then, of how far from a just society Britain has strayed that so many groups feel compelled to resort to attempts to generate compassion, which has been described as an inegalitarian, condescending, sentimental and fragile emotion (Garber, 2004). Attempts to nurture compassion should be viewed as a last-ditch recourse to unreliable and risky tactics of activism in the face of the failure to properly secure the welfare of the most vulnerable through appropriate laws and policies. An unjust society can perfectly well be highly compassionate, but we would not settle for this as a satisfactory state of affairs. In a just society, on the other hand, one could argue that individual compassion would become unnecessary. It is far preferable, Nussbaum argues, to rely 'on appropriately informed political institutions than on the vicissitudes of personal emotion' (Nussbaum, 2001, p. 387, following Immanuel Kant). The proliferation of risky attempts to nurture compassion should consequently, first and foremost, be taken as a barometer of how far the current administration of border control falls from a just system that is accountable in other, more reliable, ways.

Mitigating Considerations

The objections to compassion as a basis for social change that Nussbaum (2001) sets out in general terms and that I elaborated upon in the previous section are compelling. Nurturing or seeking compassion among the powerful must be viewed as a problematic goal for activists. This said, in this section I set out a series of mitigating considerations that might constitute grounds to view compassion somewhat more positively in specific circumstances. I do so mindful of the tendency that Berlant (2004, pp. 9–10) has identified for highly intellectual approaches to the positive emotions such as love, kindness and compassion to 'not connect, sympathize, or recognize an obligation to the sufferer ... to snuff or drown it out with pedantically shaped phrases, [and to] turn away quickly and harshly'. Here I argue that although we must view compassion sceptically, there are sometimes redeeming features of compassion that should be taken into account in doing so.

Compassion can, arguably, be highly effective in certain situations. The police officer discussed in the previous chapter who was ashamed of his work and frustrated with the laws that 'make no sense' and that lead asylum seekers into destitution had developed a keen sense of morality concerning his work. He outlined the ways in which he would assist the asylum seekers he came into contact with, helping them to complete their legal cases for support, explaining the legal process and facilitating communication between asylum seekers and local support groups. A number of officers, he told me, shared close personal relationships with asylum seekers in the local area, not only visiting them during the course of their patrols but also meeting them socially on a regular basis. This produced a high degree of loyalty towards more vulnerable migrants that sometimes meant he was willing to take personal risks of their behalf. When he became aware of information that might aid the deportation of refused asylum seekers, for instance, he often retained it rather than passing it on to enforcement officers. 'I'm not the immigration service', he told me;[25]

I'll speak to the immigration service; I mean I'm quite frank with the immigration service about what I do. But I wouldn't promote anybody being arrested, I wouldn't facilitate the arrest of somebody unless it was something serious, I mean if they were a suspect in an investigation. When I speak to them I don't say 'Did you know so and so is now living here?'

He also displayed willingness to put the needs of the asylum-seeking communities under police jurisdiction before the political pressure to meet deportation targets. 'If immigration suddenly decided they were going to go on a swoop and arrest a lot of [national group], I'll ring up [name] and say "You should be aware of this"', he explained, 'not because I'm tipping him off, errm, but because it will have a significant impact within that community'.

During my research I came across various other functionaries who are critical of what they see as unfair elements of the immigration system and act humanely whenever they can. One of the primary challenges facing detainee visiting groups is the movement of detainees from one centre to another, which can sever their contact with them, as discussed in Chapter Five. One visiting group coordinator told me of a switchboard operator who works within a detention centre and who puts visitors back in touch with detainees wherever possible when they have been moved, thereby circum-navigating the official, fax-based process of tracing detainees who have been moved, which is very time consuming. Other interviewees told me about judges who take a compassionate view of cases they hear wherever they can, and security staff who go beyond the terms of their contracts to make detainees feel welcomed. Although it is easy to dismiss these actions as fiddling while Rome burns, they meant a great deal to the migrants who benefitted from them at the time.

Individual functionaries who question border control mechanisms and logics can also send a hugely powerful message, not just to 'the public' in general as in the case of whistleblowers like Louise Perrett (mentioned in Chapter One), but also to fellow employees. Among members of the Public and Commercial Services (PCS) Union (the union that represents many employees at Lunar House), for example, there is continuing support for both the Still Human Still Here campaign, which aims to eradicate the causes of destitution of refused asylum seekers, as well as a campaign called 'Dignity for asylum seekers', which addresses difficulties with the way payments of asylum seekers' welfare benefits are made (Public and Commercial Services Union, 2014). When co-workers see their colleagues lending support to such campaigns, or simply doing their job more conscientiously and humanely, they are far more likely to reassess their own relationship to the asylum seekers under their authority.

A further mitigating factor in favour of compassion concerns the fact that efforts to nurture compassion can generate momentum towards wider, system level changes in the way immigration control is managed. In response to Nussbaum's criticisms of compassion, the distinction that she employs between individual compassion on the one hand and laws and policies on the other may not be clear cut: there may be situations in which the former can be scaled up to the latter. For the SLC activists, for example, the report into Lunar House was not the end of the story. In 2006, SLC asked twelve impartial Commissioners to conduct an independent, nation-wide review of the UK's asylum system. The Independent Asylum Commission spent two years gathering testimony from asylum seekers and the public, taking evidence from experts, and engaging in dialogue with the authorities. The Commission produced over 180 recommendations to safeguard people who seek sanctuary [in the UK], while restoring public confidence in the UK's role as a place of sanctuary for those fleeing persecution.

Citizens for Sanctuary website homepage, 2015

The Independent Asylum Commission produced the most fundamental and wide-ranging independent review of the British asylum system ever conducted (Independent Asylum Commission, 2008a, 2008b), and the campaign to implement its recommendations was ongoing at the time of writing (Citizens for Sanctuary, 2015). The Commission even introduced a new lexicon around 'sanctuary seeking' that rejected the worst connotations of the discourse around asylum seekers. This potential to scale up from the individual encounter between Mary Apragas and Mrs Lin Homer to a system-wide, political movement should not be overlooked. Although it is impossible to determine whether the risks that SLC ran in their campaign were 'worth' the eventual gains, it is helpful to think about the political poten-tial of morally obligating encounters and the intertwinement of personal ethics and public politics.

Protecting the right to act conscientiously is important too. In the United States, one American Catholic bishop has gone as far as to call for legal pro-tection for individual immigration officers who refuse to carry out tasks that they consider immoral. Bishop Thomas J. Tobin of Rhode Island wrote to the American immigration enforcement authorities in 2008 citing the legal doctrine of conscientious objection – that is, legally enshrined protection for those abstaining from orders that they view as immoral – as a way to mount a defence of immigration enforcement officers who refuse to enter churches or other sensitive sites in order to carry out deportation raids (Roman Catholic Diocese of Providence, 2008).

On the basis of these considerations it is possible to identify a set of factors that, although not enough to recommend compassion per se, are enough to mitigate towards a less unfavourable judgement in cer-tain circumstances. First, if compassion can be pursued in general terms in reference to all migrants, without relying upon particular markers such as those related to gender, age and disability, this is less likely to pro-duce the notion of the 'deserving' migrant and its unpalatable opposite. Second, if compassion can be generated not on the basis of pity, but on the basis of solidarity and concern that is not condescending, this may recommend it further. Third, if compassion at the level of individual encounters leads to a significant scaling up towards system-level change then some may judge the risks to be worth the gains. And fourth, if com-passion can be sought whilst simultaneously remaining oppositional to the overall system of controls, this may also weigh in its favour. These are, nevertheless, extremely demanding considerations.

Conclusion: The Recourse to Compassion

In this chapter I have identified and assessed the attempts of progressive asylum seeker support groups to nurture compassion among decision makers. Working through various examples, and distinguishing various scales of activity that work towards the overall objective of nurturing compassion, I have identified what sets attempts to nurture compassion apart from other forms of resistance, and I have outlined both the benefits and risks of this sort of work. I have argued that compassion needs to be approached extremely cautiously because it carries with it risks of co-optation as well as the risk of ratifying the system of immigration controls that it purportedly resists. This said, it is possible to identify a series of mitigating considerations that might cause us to look less critically upon compassion in certain circumstances.

Attempts to nurture compassion by no means describe all of the activities undertaken by activists in support of asylum seekers. Many focus exclusively upon asylum seekers themselves, providing welfare assistance, legal advice, spiritual support and social support, and do not concern themselves with decision makers or 'the public'. Others focus on direct confrontation with decision makers and bringing political pressure to bear on them without entreating them or supplicating them. Others may combine elements of many of these different tactics. But the range and number of measures designed to nurture compassion that I have outlined is indicative of the degree to which compassion-seeking has taken hold among groups working with asylum seekers in the United Kingdom. Many realise how problematic this work is, but feel as if this is the only realistic approach in the absence of more influence over the political system that has produced the current configuration of immigration controls. Indeed, it is this sense of helplessness and the unaccountability of the system of controls in Britain that is the real problem.

Notes

1 Interview with activist, October 2006.
2 Interview with charity worker, June 2011.
3 Interview with charity worker, June 2011.
4 Interview with charity worker, June 2011.
5 Interview with employee of refugee supporting charity, 26 September 2011.
6 Interview with charity worker, June 2011.
7 Interview with charity worker, June 2011.

8   Interview with refused asylum seeker, 14 July 2006.
9   Interview with refugee and activist with religious asylum support group, 13 June 2011.
10   Interview with worker at an organisation set up to offer support to Tamils, Somalis and Afro-Caribbeans living in London by refugee women, 16 June 2011.
11   Interview with refugee support group manager, 14 June 2011.
12   If a decision on an asylum seeker's claim has not been reached within 12 months through what is considered to be no fault of their own they have, since 2005, been permitted to take a job. In 2010 the stipulations were added that the job they take must be one listed on the Shortage Occupation List of the United Kingdom (Home Office, 2014a), that they cannot be self-employed, and that they cannot start a business. In practice these requirements are highly restrictive and most asylum seekers are either not allowed to work or find it extremely difficult to secure paid employment.
13   It used to be the case that if asylum seekers did not apply for asylum within three working days then they were automatically refused support under Section 55 of the Nationality, Immigration and Asylum Act (2002). Following a protracted legal battle a High Court judge ruled that this was unlawfully denying humanitarian help to those who may need it (*The Secretary of State for the Home Department vs. Wayoka Limbuela and Binyam Tefera Tesema and Yusif Adam*, 2004). Therefore, since June 2004, asylum seekers who apply for support after three working days but who do not have alternative means of support can no longer be refused emergency support (see also Home Office, 2011b).
14   Diary entry after planning meeting of Living Ghosts (South West), 16 November 2005.
15   Interview with charity manager, 10 December 2012.
16   This evidence is taken from the written submission by Freedom from Torture and Survivors Speak Out to the Home Affairs Select Committee on Asylum (Home Affairs Select Committee, 2013b).
17   Interview with former intelligence officer, 28 November 2013.
18   In 2006, for example, the UNHCR undertook a 'tremendous amount of work' (UNHCR, 2007, p. 10) designing, and supporting the UKBA to roll-out, a 55-day comprehensive training course for case workers. At the time of its inception, the UNHCR was guardedly optimistic that this course would ensure at least some consistency and fairness in the process of claim determinations. Since the course was designed, however, the training was reduced from 55 days to 25 days, resulting in the UKBA Inspectorate (a body charged with the task of inspecting and reporting on UKBA operations) reporting 'concern, particularly from managers and senior caseworkers, that the shortened module had not prepared Case Owners adequately' (Independent Chief Inspector of the UK Border Agency, 2009, p. 23). This concern is corroborated by Webber (2012) who sees the 25-day training programme as 'absurdly short' (Webber, 2012, p. 54) in the light of the wide requirements to 'learn relevant legislation and case law, interviewing, assessment of evidence and reasoning decisions' (Webber, 2012, p. 54).
19   Interview with charity worker, June 2011.
20   Interview with charity worker, June 2011.
21   Interview with detainee supporting charity worker, 1 September 2011.

22 It transpired later in the interview that this criminality largely involved minor infringements of immigration law.
23 Interview with representative from detention-focused support group, 16 September 2011.
24 Interview with charity worker, June 2011.
25 Interview with police officer, 14 July 2006.

Chapter Eight
Conclusion

My humanity is bound up in yours, for we can only be human together.

Archbishop Desmond Tutu[1]

There are no nations! There is only humanity. And if we don't come to understand that right soon, there will be no nations, because there will be no humanity.

Isaac Asimov[2]

Back at the protest outside Lunar House the English Volunteer Force (EVF) take up their position opposite us in a penned-off section on the other side of the courtyard. It is as if we are arguing over Lunar House itself, whose front sign is mounted squarely between the two sides (see Figure 8.1). Very few of them are visible. We can make out some faces but only just; mostly what we can see are police officers who have lined up between the two enclosures. The sun is beating down and some of the lead chanters need to take a break. We are in an endurance contest to keep our voices louder than theirs. The Unite Against Fascism (UAF) position on the EVF and the British National Party (BNP) is a 'no platform' position,[3] and we live that out bodily here, literally occupying the air with our sound to prevent their chanting from reaching our ears.

The UAF and the Public and Commercial Services (PCS) Union, however, were not the only organisations who voiced their opposition to the

Nothing Personal?: Geographies of Governing and Activism in the British Asylum System,
First Edition. Nick Gill.
© 2016 John Wiley & Sons, Ltd. Published 2016 by John Wiley & Sons, Ltd.

Figure 8.1   Facing the English Volunteer Force (EVF) outside Lunar House. Author's photograph, 2013.

EVF's march on Lunar House in July 2013. A few hundred metres down the road, and surrounded by a small army of police, a group of activists including members of No Borders and No One is Illegal formed their own, separate, counter-demonstration (see Figure 8.2). They refused to stand next to what they described as the 'thugs in uniform [and] the thugs in suits'[4] who worked for the United Kingdom Border Agency (UKBA), even in order to demonstrate against the common enemy of the EVF.

Before the march a heated debate about the counter protest played out through their email discussions because a minority of the No Borders and No One is Illegal group of activists wanted to send a delegation to the UAF and PCS counter protest in order to stir up radical feelings among immigration staff. Some No Borders activists were appalled at the idea of cooperating with immigration staff, calling it 'a betrayal to all those who have suffered directly or indirectly as a result of [UKBA's] work',[5] and listing the names of 20 individuals who had 'lost their lives at the hands of UKBA staff or hired thugs'.[6] They continued:

> As with fascism, the state's apparatus of immigration control: Illegalises an entire population based on arbitrary laws and perceived otherness; Keeps this population in constant fear of discovery, disqualifying them from access to the basics of survival available to others, such as work, benefits, emergency shelter;

Figure 8.2 No Borders, No One is Illegal and other groups stage a demonstration against the English Volunteer Force (EVF). Note: Lunar House is a few hundred metres to the left of the picture. Author's photograph, 2013.

Tears apart families, friends and lovers; Reduces human lives to a chain of numbers; Maintains a system of surveillance and reporting, with arbitrary and serious penalties for non-compliance; Indefinitely incarcerates 3300 people at any one time, including children, in its many detention facilities; Relentlessly collectively expels unwanted people; Inflicts widespread physical and psychological harm, and sometimes death upon these communities; Naturalises its practices to the extent that few question its legitimacy. The UK Border Agency cannot be reformed.[7]

Nevertheless, and not without a lot of soul searching, a small delegation did strike out from No One is Illegal to leaflet the immigration officers. They distributed a message among the UAF and PCS protestors that adopted a more reconciliatory tone. 'We are here today to oppose the fascists of the English Volunteer Force', it read,

But we must not forget that a bigger threat to migrants and asylum seekers is what goes on in this building [meaning Lunar House] and the buildings round about. This is where raids on … homes and workplaces are organised. This is where they organise deportations. People walk in here and their families never see them again. People are interviewed here in a traumatised state and what they say is taken down and used as evidence against them. Data collected here is used to stop people getting benefits, access to health, a proper job and so on. This is where they decide you are illegal. ... And what about you [referring to immigration officials]? It is great that some of you working here have come out to oppose the fascists. Yet you are part of this cruel and fatal machine. Is this really where you want to be, what you want to do? If you really want to defeat racism and fascism for good then rage against this machine, turn against it, undermine it, blow the whistle on it and when you get an opportunity – leave it.[8]

These two sets of sentiments reveal a great deal about the British immigration system, as well as the dilemmas that activists face in resisting it. They summarise powerfully the many subjugating aspects of the system that I have explored in this book including the impersonal nature of the system that is a typical feature of bureaucracies, as well as the remarkable normalisation and proceduralisation of lethal and subjugating controls in a society that declares itself liberal, committed to human rights and suspicious of state-sanctioned violence. They illustrate the moral poverty of the immigration control system too: the treatment of ordinary, although often vulnerable, people like criminals and the inexcusable fear and death toll that this produces.

They also capture the acute tension between outright, wholesale and direct resistance to the system of immigration controls and more pragmatic approaches that seek to engage and 'convert the policymaker' (Feldman, 2008, p. 5). The delegation hoped that by meeting and reasoning with officials they could prompt a change of heart, illustrating an optimistic view of the potential of contact and interaction. Indeed, this book began by noting the potential of proximity. Not proximity in literal terms necessarily, but moral proximity that produces encounters with others that change values and rupture the compartments that alienate officials from their subjects. The work of various philosophers, psychologists, sociologists and geographers underscores the exciting potential that proximity holds to make new things happen, including moments of ethical awakening that promise to overcome the indifference that accompanies modern border controls. It is this proximity – this togetherness – that moved the aforementioned Mr Brimelow when faced with Nataliya. Proximity is powerful enough to embolden bureaucrats to bend, question and even rewrite the rules, and as such it is a key ethical resource in a world in which rules, numbers and systems have become valued above real human life.

A key argument of this book, however, is that there are all sorts of mechanisms that limit and contain this potential. My central concern in *Nothing Personal?* has been to account for the indifference towards suffering others that pervades immigration control in terms of both its rationale and the outlook of its functionaries and managers. Moral distance, meaning the inverse correlation between distance and moral concern, provides an important starting point. For a bureaucracy to function it requires the subordination of the personal morality of bureaucrats to system objectives, and when bureaucracies are organised in ways that distance officials from subjects, either literally or institutionally, thereby making this subordination easier, there will be scant resistance from system managers. As a result of the reorganisation of border controls, including their 'upward', 'downward' and 'outward' rescaling, functionaries and subjects are kept apart in various ways as moral distance is opened.

The influence of moral distance is even more striking in light of the distancing of not only subjects from officials, but also functionaries and system managers from their subjects. 'Distancing' in the context of moral distance could easily be taken to refer to the distancing of subjects, but in Chapter Three I demonstrated that an important consequence of recent changes to the organisation of border work in the United Kingdom was the insulation, buffering and subsequent competition between back office workers who, as a result of these mechanisms, were denied a clear view of the moral consequences of their own work. This effect constitutes an important secondary form of moral distancing that alienates and estranges decision makers from the asylum seekers they make life-changing decisions about on a daily basis.

Moral distance can only go so far, however, in accounting for the indifference that riddles British border control systems. I began the book with an account of Mr Dvorzac's death in immigration detention, which cannot be explained in terms of moral distance because Mr Dvorzac was not distant from the functionaries who were responsible for him either literally or institutionally. It is clear, in this light, that further reflection on indifference is necessary, and I began this process in Chapter Four by discussing the institutional features of key meetings between decision makers and asylum seekers in the United Kingdom that preclude meaningful contact and interaction, focusing on the screening and substantive asylum interviews as well as the first instance tribunal appeal hearing. Here I departed from the most common approaches of geographers and others writing about meaningful contact, because instead of enquiring after the conditions that make meaningful contact possible, the emphasis in this chapter was on the ways in which meaningful interaction, and in particular morally demanding encounters, are suspended, avoided and averted in institutional and bureaucratic settings. It transpires that both interviews and appeals violate the majority of the conditions necessary for meaningful interaction to occur, underscoring the ways in which bureaucracies can nurture indifference even in situations of close physical proximity.

This theme was continued in Chapter Five, which focused on the ways indifference is nurtured not only in close proximity but also under the conditions of sustained contact that immigration detention produces. I used this Chapter to demonstrate that indifference is of two distinct types, corresponding to the well-known phenomenon of moral distance, on the one hand, and Simmel's observations concerning blasé attitudes towards needy others at close quarters, on the other. Whereas moral distance relies upon different forms of distance in order to be maintained – literal, organisational and temporal distance, for example – the relationship of Simmel's blasé form of indifference with distance is more complex. It is actually over-closeness, nearness to suffering and overfamiliarity with the gruesome realities of abjection that precipitates this form of indifference. In this sense this second form of indifference shares an opposite relationship with distance to the first form. Over-closeness prompts a psychological withdrawal, a recoil, that can be described as volitional psychological distancing as a form of self-care.

Chapter Six developed a further critique of theorists of moral distance and indifference by demonstrating the compatibility of emotions and subjugation. Key to this argument is the recognition that indifference can be keenly felt but conflicted in character, which distinguishes it from insensitivity. I set out evidence that many functionaries involved in border work in the United Kingdom are indifferent rather than insensitive to the suffering of their subjects. This implies that it is perfectly possible to feel shame about one's work, but to allow counter-posed emotions such as anxiety to overrule such sentiments. It also implies that it is possible to allow softer emotions such as care, empathy and even compassion into the discourse of border control, and the chapter discussed various examples and consequences of this recent paradoxical development.

This opens up the question of whether, and how, to pursue compassion among functionaries. I regret to report that the delegation from No One is Illegal to reach out to the immigration officers taking part in the UAF and PCS counter-protest was unsuccessful. For the most part the immigration officers in front of Lunar House ignored them, to the extent that, as we walked away from the march and I asked the delegation how successful their activities had been, they expressed their bitterest disappointment at the fact that so few immigration officials had engaged with them at all. Even if they had been able to interact with more officers, however, it is not at all clear that canvassing them is the best way to resist immigration controls. Although the systems and effects I have described in Chapters Two to Five tend to cordon and dampen compassion among functionaries, it does not necessarily follow that the way to respond to these effects is to seek to nurture or reinstate compassion. In fact, there are reasons to suspect that, once compassion has been limited or lost, the price of re-establishing it again may be very high in terms of the risk of portraying the 'pitiful' or 'deserving' migrant to functionaries, and the public more broadly. This approach implicitly ratifies their authority by entreating them, thereby undermining other forms of activism that question and challenge that authority.

Encounter-Aversion

These arguments hold a series of implications for the way we think about encounters, the bureaucratic management of international borders and the ways we might go about opposing subjugating structures of control. They also raise the question of whether anything general can be said about indifference beyond the case of immigration control that I have dealt with in this book.

Nothing Personal? has described the phenomenon of the encounter-aversion of bureaucracies. Positive, meaningful interaction is difficult to stage at the best of times, and morally demanding encounters even more so. Importantly, the point at which such encounters are likely to occur is not the point at which familiarity with suffering is at its highest. At this point the suffering threatens to overwhelm moral sentiments (Hamblet, 2003). This is a phenomenon sometimes overlooked by Bauman in his theorisation of moral distance, resulting in an almost naïve faith in the possibilities of proximity. In order to avoid this mistake we might think instead about the existence of an optimal frequency and intensity of contact – close but not too close – that affords the possibility of concern. We might expect this frequency to be maintained by particular and specific relationships with others, rather than being forced to treat subjects as a group, population or a 'throughput'. Unfortunately, because this optimal point can be disturbed *both* by increasing *and* decreasing the frequency and intensity of contact, it constitutes a rather unstable equilibrium. Contrary to the robustness of indifference, it is fragile: all the more so because of the bureaucratic tendency to both under- and over-stimulate the empathetic sensibilities of decision makers.

Encounter-aversion is not, on the whole, the result of the calculated and premeditated plotting and planning of a sinister, monstrous state, but the result of the tendency of bureaucratic systems to take the paths of least resistance available to them in the course of their own operation. Although on the surface this might sound like good news, the lack of a calculative and identifiable 'centre' arguably strengthens the system itself. For John Law,

> domination is often not a system effect, the consequence of a coherent order. Rather it is a result of non-coherence. Of elements of structuring, ordering, that only partially hang together. Of relations of subordination that are relatively invulnerable precisely because they are not tightly connected. Invulnerable because when one is pulled down the others are not pulled down with it.
>
> Law, 2008, p. 641

The tensility that non-coherence affords to immigration control systems is bolstered by the almost permanent sense of crisis that commentators, especially in the press, associate with immigration control in Britain. Perversely, this sense of crisis improves systemic flexibility and agility, precisely because systems are permanently viewed as in need of reform, rendering change expected and acceptable (see Klein, 2007; Mountz and Hiemstra, 2014). In the United Kingdom 'public consent has been procured' (Tyler, 2013, p. 211) in this way for a dizzying pace of legislative and policy churn over the past 20 years that I set out in Chapter One. My preference for describing the effects I have explored in this book as bureaucratic tendencies rather than state strategies reflects these properties of the system of border controls.

The tendency to remove, or efface, the individual subject that I have described in *Nothing Personal?* is not as problematic in other situations in which bureaucratic modes of organisation are employed. For Du Gay (2000), the achievement of bureaucracy is its ability to treat people equally and according to similar standards, thus doing justice to their often competing or unrealistic claims in a morally defensible way. This imperative, he argues, outweighs detractors' concerns that bureaucracy is coldly impassive and insensitive. In the face of intractable dilemmas over how to distribute scarce resources this argument seems plausible. But I hold asylum seeking, and the demands of migrants in general, to be distinct from such dilemmas of government because there is no *a priori* dilemma in relation to asylum seekers that is not created by the existence of international borders themselves: the problem of managing international borders is self-inflicted by the system of nation states. The media-fuelled anxieties about being 'swamped' or 'flooded' by migrants generally remain trapped in nation-based ways of thinking that lack the vision to appreciate how humankind might benefit from a lack of international borders (Hayter, 2004; Gill, 2009; Anderson *et al.*, 2009). In this light, assessing the effects of the bureaucratic management of border control is not a matter of balancing the need for a just and impartial approach to the intractable issues thrown up by border management against the slowness, red tape and frustrating impassiveness of bureaucratic control, because the very notion of just and fair international borders is one that seems incredibly strange to accept. International borders help, immeasurably, to keep the global poor poor and the global rich rich, they randomly assign life chances and opportunities on the basis of the lottery of birth, and they support an exploitative division of global labour without which globalisation itself would be severely impoverished (Bayart, 2007). In the asylum context, borders also help to keep the consequences of violence that can often be traced to histories of colonial exploitation away from the historical colonisers. These are not the results of 'mismanaged', 'unjust' or 'unfair' international borders, these are the results of international borders per se. It is on this basis – which I admit is an ideological one rather than a practical or worldly approach that will be of much immediate use to policy makers involved in border control – that I feel at liberty to critique the consequences of the bureaucratic management of border control without conceding its benefits.

Remaining Oppositional

Therefore, if readers had expected me to provide a list of reforms here, directed at government, that would allow a shift away from an unjust system of immigration controls towards a more just set of controls, they will be disappointed. Nevertheless, I am conscious that I have painted a bleak picture of immigration controls and, through my critical discussion of compassion, a bleak picture of the prospect of resisting them as well. It follows from my comments in Chapter Seven about compassion and the inherent difficulties of softening and humanising the immigration control system, however, that the most effective forms of resistance would not distinguish between deserving and undeserving migrants, and would instead insist upon free international movement for all, regardless of whether migrants have something to offer to their destination countries and regardless of their background, skill set or history. According to this reasoning the clearest and most logically consistent argument for change comes from radical quarters: No Borders and No One is Illegal, who campaign for an end to international immigration controls in general. As the No One is Illegal Manifesto puts it:

> We are opposed to all arguments that seek to justify the presence of anyone on the grounds of the economic or cultural or any other contributions they may make. It is not up to the British state to decide where people should or should not live, or anyone else but migrants and refugees themselves. We support the unfettered right of entry of the feckless, the unemployable and the uncultured. We assert No One Is Illegal.
>
> No One is Illegal, 2003

My prescription to Western governments who are busy building ever-taller barriers to international immigration is to abandon this activity and begin to undertake the (admittedly complex) process of dismantling them. Limiting the liberty of others, including their international mobility, is a *punishment* for which, for the vast majority of those subject to immigration controls, no crime has been committed. Perhaps there is a case to retain some measure of control when it comes to proven terrorists and proven, serious criminals because, in these cases, punishment may be justified and limiting international mobility may be necessary to protect innocent people from the risk of malevolence. But outside these cases, immigration controls embody a form of institutionalised injustice. Rather than suggest a series of reforms then, I prefer to pose a question: who among our governments has the courage and political skill to lead the way towards a different global future without international immigration controls? Thinking in these terms promises to puncture the suffocating consensus that we need international border controls,[9] and that they need to be firmer and tighter: a consensus that currently characterises the rather technical and under-ambitious public debates that tend to occur around British border controls.

No Borders and No One is Illegal reject the objective of softening and humanising the system of international immigration controls. For them, there is no such thing as a humane international immigration control system and it is more unrealistic to expect to be able to make international immigration controls just and fair than it is to expect them to be completely disbanded, because at least the latter is possible.

> The demand for no controls – based on the assertion that no one is illegal – is frequently derided as utopian and is compared adversely to the 'realism' of arguing for fair controls. However this stands political reality on its head. The struggle against the totality of controls is certainly uphill – it may well require a revolution. However the achievement of fair immigration restrictions – that is the transformation of immigration controls into their opposite – would require a miracle.
>
> No One is Illegal, 2003

For them, there is nothing that can be improved about the system of immigration controls to make up for the fact that it is arbitrarily unjust and imperialist in legacy. These groups draw inspiration from anarchist principles to question the need for one of the most pervasive and exploitative state interventions and checks over individual freedom that is in force in the world today: the limitation on free international movement.

The trouble with utopian ideas such as the dissolution of international border controls is that they can be dismissed as unrealistic by 'men of state' who think in terms of the existing status quo, and who cannot think beyond it (Lefebvre, 2009, p. 54).[10] But not only do utopian ideas serve the function of providing a navigational aide through the messiness of everyday skirmishes and bargains around border controls. They also, in the words of David Harvey (2000, p. 281), 'hold up for inspection the waste and foolishness of [our] times', and, in so doing, provide the basis upon which we can 'insist that things could and must be better' (Harvey, 2000, p. 281).

This is not to say that nothing can be done by oppositional groups in the absence of a wholesale dismantling of border controls. On the contrary, it is perfectly possible to remain oppositional and defiant even in situations where revolution is not yet feasible and the system being resisted is unreformable. This was de Certeau's (1984) point when he described the tactics of the subjugated: surreptitious, agile, diversionary, agitating and subversive courses of action that are capable of chipping away at the established order without condoning it for a moment.[11]

Generalised Indifference

Whereas I have discussed indifference as it relates to border control, various commentators have made the case that indifference to suffering others is a general feature of modern life. For example, in 2006 Barack Obama remarked that:

There's a lot of talk in this country [referring to the United States] about the federal deficit. But I think we should talk more about the empathy deficit – our ability to put ourselves in someone else's shoes, to see the world through those who are different from us – the child who's hungry, the laid-off steelworker, the immigrant cleaning your dorm room. … We live in a culture that discourages empathy, a culture that too often tells us that our principal goal in life is to be rich, thin, young, famous, safe and entertained.

Barack Obama, Speech to the Graduates of Northwestern University in 2006[12]

And when His Holiness Pope Francis I visited the small Italian island of Lampedusa, the site of horrific migrant deaths in 2013,[13] he connected migrant deaths with profound general shortcomings in modern moral culture:

> The culture of well-being, that makes us think of ourselves, that makes us insensitive to the cries of others, that makes us live in soap bubbles, that are beautiful but are nothing, are illusions of futility, of the transient, that brings indifference to others, that brings even the globalization of indifference. In this world of globalization we have fallen into a globalization of indifference. We are accustomed to the suffering of others, it doesn't concern us, it's none of our business.

His Holiness Pope Francis I, 8 July 2013, Lampedusa[14]

From this viewpoint indifference to suffering others can be viewed as a global moral pandemic: the logical consequence of capitalism's relentlessly individuating culture, brutal media tactics and bureaucracy's depersonalising effects.

The agenda that I would recommend for the coming years to geographers and other social scientists with an interest in the suffering of others would consequently place indifference centre stage. We need to understand the dynamics of indifference more fully, in terms of its relation to specifically geographical phenomena such as separation, avoidance and different forms of distance. What I have set out in *Nothing Personal?* is a starting point, but it raises questions about the extent to which the analysis of the relation between distance and indifference that I have undertaken is generalisable to other contexts of suffering and vulnerability, whether there are exceptions, and what factors might disturb it. Underpinning all of these questions should be a focus on the interpersonal – the level of interaction and relations between two or a small number of people or beings. It is at this level that sociologists, moral philosophers, geographers and psychologists have located morality, and therefore it is the factors that shape, constrain and facilitate interpersonal relationships that should be of interest to an investigation into the circumstances of moral failure.

Notes

1   Taken from Tutu (1989), page 69.
2   Taken from Asimov (1994), page 421.

3   A 'no platform' position refers to the belief that certain opposition groups, such as those that are understood to be espousing fascist ideas, should not be allowed a public platform or to enter into public debate.

4   This quote is taken from an email circulated on 16 July 2013 around the email listserv migrationstruggles@lists.riseup.net and was signed 'No Borders London'.

5   This quote is taken from an email circulated on 16 July 2013 around the email listserv migrationstruggles@lists.riseup.net and was signed 'No Borders London'.

6   This quote is taken from an email circulated on 16 July 2013 around the email listserv migrationstruggles@lists.riseup.net and was signed 'No Borders London'.

7   This quote is taken from an email circulated on 16 July 2013 around the email listserv migrationstruggles@lists.riseup.net and was signed 'No Borders London'.

8   Leaflet distributed at the counter-protest against the EVF outside Lunar House, 27 July 2013.

9   I recognise that the dissolution of international border controls would not, on its own, solve all of the world's ills; see Gill (2009) for a fuller discussion.

10 Lefebvre (2009, p. 54) describes men of state as people who 'accept the existing State as a central given of reality, as a central given of the moral sciences, who think as a function of this given and who pose all the problems related to the knowledge of society, to science and reality itself as a function of this given'.

11 See also Scott (1985) and, for a discussion specific to immigration control in the United Kingdom and United States, Gill *et al.* (2014).

12 Quote taken from http://www.northwestern.edu/newscenter/stories/2006/06/barack.html, accessed 4 August 2015.

13 On 3 October 2013 a boat carrying migrants from Libya to the small Italian island of Lampedusa sank after its engine failed and a fire broke out on board. The boat was grossly overcrowded and ill-equipped for emergencies, and the death toll exceeded 360, eliciting widespread expressions of shock and concern from the international community.

14 Quote taken from http://en.radiovaticana.va/news/2013/07/08/pope_on_lampedusa:_%E2%80%9Cthe_globalization_of_indifference%E2%80%9D/en1-708541, accessed 4 August 2015.

Methodological Appendix

Nothing Personal? draws on a series of four funded research projects that constitute a programme of research that began in 2003 and is ongoing at the time of writing. The first project was doctoral research funded by the Economic and Social Research Council (ESRC)[1] at the University of Bristol, UK, that aimed to explore the challenges facing immigration decision makers in the United Kingdom from the perspective of both decision makers themselves and asylum seekers in their care (henceforth the 'decision makers' project). I worked on this project alone, although ably supervised by Professors Adam Tickell and Wendy Larner, and conducted interviews and participant observation in person. The second project was also funded by the ESRC[2] and examined the challenges and mitigating strategies of asylum support groups in the United Kingdom and United States (henceforth the 'asylum support groups' project), and for this project I was assisted by two co-investigators, Drs Deirdre Conlon and Imogen Tyler, as well as a researcher, Dr Ceri Oeppen. The third project, which is ongoing at the time of writing and again funded by the ESRC,[3] examines the asylum appeal process (specifically the first tier) in England and Wales (henceforth the 'tribunals' project). For this project I have been assisted by Ms Jennifer Allsopp and Drs Melanie Griffiths, Andrew Burridge, Rebecca Rotter and Natalia Paszkiewicz. The fourth project is actually a series of seminars, also ongoing at the time of writing, funded by the ESRC and on the theme of immigration detention,[4] which has given rise to many discussions and presentations that have informed my argument in this book.

Although the projects form a coherent programme of research they also span a variety of different methodologies and approaches, many of which

Nothing Personal?: Geographies of Governing and Activism in the British Asylum System, First Edition. Nick Gill.

I have drawn on in *Nothing Personal?*. They have entailed focus groups, surveys, interviews, ethnography and document analysis. To explore all the methodological challenges and dilemmas I have faced, and decisions that I have taken, for each of these projects and approaches would require a much fuller treatment than I have space for here, so I will focus upon a subset of methodological issues including those relating to sampling, access, ethics and my approach to analysing the data.

Sampling and Access

For the decision makers project and the tribunals project, a key issue was how to access secure places of immigration control. In the case of the tribunals this is relatively easy to address: it is possible in England and Wales to attend immigration hearings by sitting in the public area of the hearing rooms, and over the course of the project around 400 appeals have been observed in this way. For the decision makers project, access was more challenging. I required access to officials in Lunar House, as well as the National Asylum Support Service's (NASS's) back offices in Portishead and Campsfield House immigration removal centre. My strategy involved writing letters to gatekeepers in these locations, and although I had to be persistent (such as sending multiple letters to various individuals, and following up with telephone calls frequently in order to get past secretaries and be allowed to talk to the decision makers themselves) this strategy did eventually succeed. I was allowed access to the NASS offices including being shown around and introduced to a variety of staff there. I was allowed to interview the staff at the office and snowballed from established contacts to new ones. I was also granted entry to Campsfield House immigration removal centre, where I was given a tour (although I was always accompanied) and some detailed interviews with the staff working at the centre. And in Croydon I was allowed into the management offices (only once, under supervision), to interview a senior immigration decision maker. The key limitations to my access included not gaining permission to talk to existing detainees (I had to interview ex-detainees, although at least then we were not accompanied by an official and they could talk more openly) and not having the opportunity to conduct a placement at the NASS offices (I had to satisfy myself with interviews and being shown around). On the whole, however, I was surprised by the willingness of staff to talk to me and give their time generously in support of my studies.

Access to asylum support group workers during the asylum support group project was far less challenging and I found voluntary and charitable sector workers in particular generous and approachable even though I was asking to interview them at a time of funding cuts and consequently great pressure in their sector. One interviewee memorably told Ceri that he lost his job the same week as the interview.

In terms of sampling, although I was not in pursuit of a representative sample in a strictly statistical sense, I tried to balance various factors in terms of both the choice of research sites and research participants. During the decision makers project I sought to sample across the sites at which the various stages of the asylum claim determination process took place. This meant researching at Lunar House where initial claims are lodged, at NASS offices where paperwork relating to claims for welfare support is dealt with, and at Campsfield where asylum seekers are confined. During the asylum support group project we attempted to balance between different regions of the United Kingdom and United States, and also to represent different types of support group, including small, informal groups as well as well-established charities. During the tribunals project, especially during the ethnographic phase, we examined different tribunals across urban and suburban contexts and with different levels of busyness.

The decision makers project entailed 39 interviews. Of those, eight refused to be recorded and two were conducted by email. Six were with either current or former asylum seekers. Where interviews were not recorded I made detailed scratch notes about the interview immediately afterwards, including any direct quotes I had noted down during the conversation. Initial contact was usually made by sending a letter and then calling via telephone, although the asylum seekers and former asylum seekers in the sample were contacted via word of mouth. Roughly equal numbers of inter-viewees were contacted cold (15) as were snowballed (16) (the remainder I met in various other contexts). Most interviews took place at the interviewee's place of work (20) but it was also common to conduct inter-views in a public place (9) or at the interviewee's home (8).

The asylum support group project drew on survey evidence with 130 groups, interviews with 35 of these and three focus groups. Of the 35 inter-viewees, 22 were US-based and are almost exclusively not drawn upon in *Nothing Personal?*, whereas the remaining 13 were with UK-based asylum support groups and feature more frequently in *Nothing Personal?* The characteristics of the survey respondents, which roughly reflect those of the interview respondents as well, included (these are non-exclusive categories): (i) 58% were involved in providing legal support; (ii) 56% were aiming to change government policy; (iii) 59% stated that the provision of food, medical care or other services was the primary aim of their work; (iv) 50% stated that allowing asylum seekers to express themselves creatively was involved in their work; and (v) 32% stated that campaigning was involved in their work. Additionally, 52% employed fewer than 10 people, 30% were motivated by faith, and 88% were motivated by their concern for human rights.

The tribunal ethnographies drew on a 3-month intensive ethnographic study of immigration tribunals in the United Kingdom. The researchers,

Andy and Mel, conducted around 100 observations of hearings during 2013, which paved the way for a further 290 more structured observations of appeal hearings in 2014. Observations were recorded in the form of detailed research diaries.

Ethics

For the decision makers project, formal consent forms were used. Where feasible these were signed by the interviewees before their interviews, and set out the uses of the data, the fact that the data would be used anonymously and their ability to withdraw from the project at any time. Both decision makers and asylum seekers, however, would sometimes refuse to sign the consent form but were prepared to conduct an interview nonetheless. In these situations I attempted to replicate the key information orally at the start of the interview.

In the case of the tribunals project the observations were of a public hearing and, after some consideration, we decided that requiring the applicant to sign a consent form would constitute an additional source of stress at an already stressful time. As a consequence I did not ask the researchers to seek the consent of the asylum applicants being observed when they thought this might have added to the anxiety of the appellant. Because the hearings are public events during which the actors can expect to be observed this approach seemed appropriate. For their part, Her Majesty's Courts and Tribunals service told us that we were at liberty to conduct the research on condition that any anonymity or reporting restrictions in place over the cases that we observed were respected. Hearings that involved particularly sensitive issues were sometimes inaccessible as they were conducted in camera at the discretion of the judge.

Maintaining the anonymity of respondents has been a high priority throughout all of the projects. The risks to individual asylum seekers of publishing details of their cases in particular are acute, especially if a final decision on their claim has not been reached. It is interesting, however, that some interviewees – activists, asylum seekers and activist asylum seekers – requested that I waive my practices regarding anonymity so as to provide exposure for their situation. In other words they preferred to not be anonymised because they wanted as many people to know about what happened to them as possible. This poses something of a dilemma to the researcher, who could potentially be very beneficial to asylum seekers if they broke with convention and included personal details in publications. Usually one wants to be as obliging as possible to the research participants. Nevertheless, on the basis that at least some risk is involved in revealing the names of participants I decided against doing so on the basis that there may be unpredictable changes in their circumstances in the future that could cause them to regret the publication of details of their situations. This was a difficult decision because it required me to not carry out the wishes of the participants.

Various other ethical considerations have presented themselves through the research but I would like to comment in particular on the issue of secondary trauma among researchers working with suffering others. Speaking personally, becoming party to the accounts of individuals struggling with traumatic experiences in their home countries, compounded by traumatic experiences in the United Kingdom, has often been challenging. I have frequently found the work exasperating and felt unable to make any sort of difference. Speaking as a manager of researchers who have exposed themselves to disturbing accounts of trauma day after day for extended periods of several months at a time, I have witnessed the stress and fatigue that this sort of work can also produce. And speaking as a researcher investigating asylum support groups, I can confirm the effect that secondary trauma can have in generating blasé and detached attitudes within supposedly supportive organisations. Taking the threat of secondary trauma seriously is a key consideration of ethical research in this area in the future, and should also be a key consideration of courts and the Home Office in protecting their judges, barristers, solicitors, translators and caseworkers.

Analysis

I analysed the data, which are a varied collection of notes, diary entries, emails, transcriptions, websites, sketches, photographs, letters, newspapers, policy documents and speadsheets, by following the principles of coding for research analysis (Crang, 1997). Although I did not use a computer package to code, I tended to define a small number of codes before reading the data, and then expand this number as I went along when certain findings seemed particularly relevant and did not fit into any of my already existing codes. Although this tended to mean that the number of codes I employed proliferated with each reading of the data, this approach allowed me to balance between seeking out the evidence that I thought was most pertinent on the one hand, and allowing room for the data to guide me towards unexpected insights on the other. This means of analysis can be understood as a crude balance between the etic (from the perspective of the researcher) and emic (from the perspective of the subject) types of data analysis that are common in the social sciences (Merriam, 1998).

Notes

1   1+3 doctoral award from 2003 to 2007, grant number PTA-030-2003-01643.
2   Grant number RES-000-22-3928-A.
3   Grant number ES/J023426/1.
4   Grant number ES/J021814/1.

References

Ahmed, S. (2004) *The Cultural Politics of Emotion*. Abingdon: Taylor & Francis.

Aitken, S. (2010) 'Throwntogetherness': encounters with difference and diversity. In D. DeLyser, S. Herbert, S. Aitken, M. Crang and L. McDowell (eds) *The SAGE Handbook of Qualitative Geography*. London: Sage, pp. 46–68.

Allen, J. (1990) Does feminism need a theory of 'The State'? In S. Watson (ed.) *Playing the State: Australian Feminist Interventions*. New York: Verso, pp. 21–37.

Allport, G. (1954) *The Nature of Prejudice*. Reading, MA: Addison-Wesley Publishing Company.

Amin, A. (2002a) Ethnicity and the multicultural city: living with diversity. *Environment and Planning A* 34(6): 959–980.

Amin, A. (2002b) *Ethnicity and the Multicultural City: Living with Diversity*. Report for the Department of Transport, Local Government and the Regions and the ESRC Cities Initiative. Durham: University of Durham. Available at http://red.pucp.edu.pe/ridei/wp-content/uploads/biblioteca/Amin_ethnicity.pdf (accessed 30 August 2015).

Amin, A. and Thrift, N. (2002) *Cities: Reimagining the Urban*. Cambridge: Polity Press.

Amnesty International (2004) *Get It Right: How Home Office Decision-Making Fails Refugees*. London: Amnesty International. Available at http://www.amnesty.org.uk/sites/default/files/get_it_right_0.pdf (accessed 29 July 2014).

Amnesty International and Still Human Still Here (2013) *A Question of Credibility: Why So Many Initial Asylum Decisions Are Overturned On Appeal in the UK*. London: Amnesty International. Available at http://www.amnesty.org.uk/sites/default/files/a_question_of_credibility_final_0.pdf (accessed 29 July 2014).

Nothing Personal?: Geographies of Governing and Activism in the British Asylum System, First Edition. Nick Gill.

© 2016 John Wiley & Sons, Ltd. Published 2016 by John Wiley & Sons, Ltd.

Amoore, L. (2009) Algorithmic war: everyday geographies of the war on terror. *Antipode* 41(1): 49–69.

Amoore, L. and Hall, A. (2009) Taking people apart: digitised dissection and the body at the border. *Environment and Planning D: Society and Space* 27(3): 444–464.

Anderson, B. (2013) *Us and Them? The Dangerous Politics of Immigration Control.* Oxford: Oxford University Press.

Anderson, B., Sharma, N. and Wright, C. (2009) Editorial: Why No Borders? *Refuge* 26(2): 5–18.

Anderson, M. and Den Boer, M. (1994) *Policing Across National Boundaries.* London: Pinter.

Andrijasevic, R. (2010) DEPORTED: The right to asylum at EU's external border of Italy and Libya. *International Migration* 48(1): 148–174.

Andrijasevic, R. and Walters, W. (2010) The International Organization for Migration and the international government of borders. *Environment and Planning D: Society and Space* 28(6): 977–999.

ANSAmed (2013) Italy-Libya: border control a priority, Letta tells Zeidan. ANSAmed news. Available at http://www.ansamed.info/ansamed/en/news/nations/libya/2013/07/04/Italy-Libya-border-control-priority-Letta-tells-Zeidan_8977604.html (accessed 13 November 2015).

Apple, M.W. (1995) *Education and Power.* New York: Routledge.

Arendt, H. (1994) Understanding and politics (the difficulties of understanding). In J. Kohn (ed.) *Essays in Understanding 1930–1954: Formation, Exile and Totalitarianism.* New York: Schocken Books, pp. 307–327.

Arendt, H. (1964/1994) *Eichmann in Jerusalem: A Report on the Banality of Evil.* New York: Penguin Books.

Aristotle (1926) *Aristotle in 23 Volumes* (trans. J.H. Freese), Vol. 22. Cambridge and London: Harvard University Press; William Heinemann Ltd.

Asimov, I. (1994) *I. Asimov.* New York: Doubleday.

Askins, K. and Pain, R. (2011) Contact zones: participation, materiality, and the messiness of interaction. *Environment and Planning D: Society and Space* 29: 803–821.

Asylum Aid (2013) *The Asylum Process Made Simple.* Available at http://www.asylumaid.org.uk/pages/the_asylum_process_made_simple.html (accessed 19 February 2013).

Athwal, H. (2014) *Deaths in Immigration Detention: 1989–2014.* Institute of Race Relations. Available at http://www.irr.org.uk/news/deaths-in-immigration-detention-1989-2014/ (accessed 19 November 2014).

Back, L. (2007) *The Art of Listening.* Oxford: Berg.

Back, L., Farrell, B. and Vandermaas, E. (2005) *A Humane Service for Global Citizens: Report on the South London Citizens Enquiry Into Service Provision by the Immigration and Nationality Directorate at Lunar House.* London: South London Citizens.

Bacon, C. (2005) The evolution of immigration detention in the UK: the involvement of private prison companies. *Refugee Studies Centre Working Paper No. 27.* Oxford: Refugee Studies Centre. Available at http://www.rsc.ox.ac.uk/publications/the-evolution-of-immigration-detention-in-the-uk-the-involvement-of-private-prison-companies (accessed 30 August 2015).

Bail for Immigration Detainees (2009) *Out of Sight, Out of Mind: Experiences of Immigration Detention in the UK.* London: Bail for Immigration Detainees.

Available at http://www.osservatoriomigranti.org/assets/files/BID%20-%20Out%
20ofsight%20out%20of%20mind.pdf (accessed 30 August 2015).

Bail for Immigration Detainees and the Information Centre about Asylum and
Refugees at the Runnymede Trust (2011) *Summary Findings of Survey of Levels of
Legal Representation for Immigration Detainees Across the UK Detention Estate
(Surveys 1–6)*. 13 September 2013. BID Research Reports.

Bailey, A.J., Wright, R.A., Miyares, I. and Mountz, A. (2002) Producing Salvadoran
transnational geographies. *Annals of the Association of American Geographers* 92(1):
125–144.

Barlow, K., Paolini, S., Pedersen, A. et al. (2012) The contact caveat: negative
contact predicts increased prejudice more than positive contact predicts reduced
prejudice. *Personality and Social Psychology Bulletin* 38(12): 1629–1643.

Barnardo's (2014) *Cedars: Two Years On*. Ilford, Essex: Barnardo's. Available at http://
www.barnardos.org.uk/16120_cedars_report.pdf (accessed 29 July 2014).

Barnett, C. (2005) Ways of relating: hospitality and the acknowledgement of otherness. *Progress in Human Geography* 29(1): 5–21.

Barrett, D. and Ensor, J. (2012) Judges who allow foreign criminals to stay in Britain.
The Telegraph, 16 June. Available at http://www.telegraph.co.uk/news/uknews/
immigration/9335689/Judges-who-allow-foreign-criminals-to-stay-in-Britain.
html (accessed 11 September 2015).

Bauder, H. (2003) Equality, justice and the problem of international borders: the
case of Canadian immigration regulation. *ACME: An International E-Journal for
Critical Geographies* 2(2): 167–182.

Bauman, Z. (1989) *Modernity and the Holocaust*. Cambridge: Polity Press.

Bauman, Z. (1993) *Postmodern Ethics*. London: Routledge.

Bauman, Z. and Donskis, L. (2013) *Moral Blindness: The Loss of Sensitivity in Liquid
Modernity*. Cambridge: Polity Press.

Bayart, J.-F. (2007) *Global Subjects: A Political Critique of Globalization*. Cambridge:
Polity Press.

BBC (2005a) Rural asylum centre plans dropped. 11 June 2005. Available at http://
news.bbc.co.uk/1/hi/uk_politics/4083032.stm (accessed 21 July 2014).

BBC (2005b) *File on Four: Immigration* [transcript]. 10 December 2006. Available
at http://news.bbc.co.uk/nol/shared/bsp/hi/pdfs/21_06_05_asylum.pdf (accessed
7 January 2008).

BBC (2005c) Detention Undercover – the Real Story. 28 February 2005. Available
at http://news.bbc.co.uk/1/hi/programmes/real_story/4304547.stm (accessed 26
June 2013).

BBC (2006a) Death in a cold, strange land. 24 March 2006. Available at http://
news.bbc.co.uk/1/hi/england/4582470.stm (accessed 29 July 2014).

BBC (2006b) Clarke insists 'I will not quit'. 25 April 2006. Available at http://news.
bbc.co.uk/1/hi/uk_politics/4944164.stm (accessed 28 July 2014).

BBC (2006c) Illegal workers prompt new probe. 19 May 2006. Available at http://
news.bbc.co.uk/2/hi/uk_news/politics/4995764.stm (accessed 29 July 2014).

Benedict, S. (2003) Caring while killing: nursing in the "euthanasia" centers, In E.
Baer and M. Goldenberg (eds) *Experience and Expression: Women, the Nazis, and the
Holocaust*. Detroit: Wayne State University Press, pp. 95–110.

Berlant, L. (ed.) (2004) *Compassion: The Culture and Politics of an Emotion*. New York &
London: Routledge.

Bethell, N. (1974) *The Last Secret: Forcible Repatriation to Russia 1944-7*. London: Andre Deutsch.

Bialasiewicz, L. (2012) Off-shoring and out-sourcing the borders of Europe: Libya and EU border-work in the Mediterranean. *Geopolitics* 17(4): 843–866.

Bigo, D. and Tsoukala, A. (eds) (2008) *Terror, Insecurity and Liberty: Illiberal Practices of Liberal Regimes after 9/11*. London: Routledge.

Bögner, D., Herlihy, J. and Brewin, C.R. (2007) Impact of sexual violence on disclosure during Home Office interviews. *British Journal of Psychiatry* 191: 75–81.

Bohmer, C. and Shuman, A. (2008) *Rejecting Refugees: Political Asylum in the 21st century*. London: Routledge.

Bondi, L. (2008) On the relational dynamics of caring: a psychotherapeutic approach to emotional and power dimensions of women's care work. *Gender, Place and Culture* 15(3): 249–265.

Boswell, C. (2003) Burden-sharing in the European Union: lessons from the German and UK experience. *Journal of Refugee Studies* 16(3): 316–335.

Bosworth, M. (2014) *Inside Immigration Detention*. Oxford: Oxford University Press.

Bousfield, D. (2005) The logic of sovereignty and the agency of the refugee: recovering the political from 'bare life". *YCISS Working Paper No. 36*. Toronto: York University. Available at http://yciss.info.yorku.ca/files/2012/06/WP36-Bousfield.pdf (accessed 29 July 2014).

Brenner, N. (2004) *New State Spaces: Urban Governance and the Rescaling of Statehood*. Oxford: Oxford University Press.

Bright, M. (2003) Welcome to Immigration Central. Please join the queue: your number is 110,001…. *The Observer*, 2 March. Available at http://www.theguardian.com/politics/2003/mar/02/immigration.immigrationpolicy (accessed 30 August 2015).

Bright, M. (2004) Asylum seekers? Not here, not even for a few minutes. *The Observer*, 25 April, p. 5. Available at http://www.theguardian.com/uk/2004/apr/25/immigration.immigrationpolicy (accessed 30 August 2015).

Briskman, L. and Cemlyn, S. (2005) Reclaiming humanity for asylum seekers: a social work response. *International Social Work* 48(6): 714–724.

British Red Cross (2010) *Not gone, but forgotten: the urgent need for a more humane asylum system*. London: British Red Cross. Available at http://www.redcross.org.uk/About-us/News/2010/June/~/media/BritishRedCross/Documents/Archive/GeneralContent/N/Destitution%20report%20Not%20gone%20but%20forgotten.ashx (accessed 7 March 2011).

Bruegel, I. and Natamba, E. (2002) Maintaining contact: what happens after detained asylum seekers get bail? *Social Science Research Paper no. 16*. London: South Bank University.

Burdsey, D. (2013) 'The foreignness is still quite visible in this town': multiculture, marginality and prejudice at the English seaside. *Patterns of Prejudice* 47(2): 95–116.

Campbell, J. (2009) Refugees and the law: an ethnography of the British asylum system. Non-technical summary (Research summary). ESRC End of Award Report *Res-062-23-0296*. Swindon: ESRC.

Casciani, D. (2004) How Portishead divided over asylum. BBC News Online, 21 April 2004. Available at http://news.bbc.co.uk/1/hi/uk/3644595.stm (accessed 28 July 2014).

Century, G., Leavey, G. and Payne, H. (2007) The experience of working with refugees: counsellors in primary care. *British Journal of Guidance and Counselling* 35: 23–40.

Chamayou, G. (2013) *Théorie du drone*. La Fabrique Editions.

Channel Four (2015) Yarl's Wood: Undercover in the Secretive Immigration Centre. Available at http://www.channel4.com/news/yarls-wood-immigration-removal-detention-centre-investigation (accessed 27 August 2015).

Chantler, K. (2010) Women seeking asylum in the UK: contesting conventions. In I. Palmary (ed.) *Gender and Migration: Feminist Interventions*. London: Zed Books, pp. 86–103.

Chatterton, P. and Pickerill, J. (2010) Everyday activism and transitions towards post-capitalist worlds. *Transactions of the Institute of British Geographers* 35(4): 475–490.

Citizens Advice (2003) Memorandum submitted by Citizens Advice. Written Evidence to the Home Affairs Select Committee Inquiry into Asylum Applications, 2003. Available at http://www.parliament.the-stationery-office.co.uk/pa/cm200304/cmselect/cmhaff/218/218we09.htm (accessed 30 July 2014).

Citizens for Sanctuary (2015) *Citizens for Sanctuary: Securing Justice for People Fleeing Persecution. Rebuilding Public Support for Sanctuary* (website homepage). London: http://www.citizensforsanctuary.org.uk/ (accessed 10 November 2015).

Clayton, G. (2010) *Immigration and Asylum Law*. Oxford: Oxford University Press.

Cohen, S. (2001) *States of Denial: Knowing about Atrocities and Suffering*. Cambridge: Polity Press.

Cohen, S. (2002) The local state of immigration controls. *Critical Social Policy* 22(3): 518–543.

Coleman, M. (2009) What counts as the politics and practice of security, and where? Devolution and immigrant insecurity after 9/11. *Annals of the Association of American Geographers* 99(5): 904–913.

Conlon, D. (2011) Waiting: feminist perspectives on the spacings/timings of migrant (im)mobility. *Gender, Place and Culture: A Journal of Feminist Geography* 18(3): 353–360.

Conlon, D. (2013) Hungering for freedom: asylum seekers' hunger strikes - rethinking resistance as counter-conduct. In D. Moran, N. Gill and D. Conlon (eds) *Carceral Spaces: Mobility and Agency in Imprisonment and Migrant Detention*. Farnham: Ashgate, pp. 133–148.

Corbridge, S. (1993) Marxisms, modernities, and moralities: Development praxis and the claims of distant strangers. *Environment and Planning D: Society and Space* 11(4): 449–472.

Crang, M. (1997) Analyzing qualitative materials. In R. Flowerdew and D. Martin (eds) *Methods in Human Geography: A Guide for Students Doing a Research Project*. Harlow: Longman, pp. 183–196.

Crawley, H. (2010) 'No-one gives you a chance to say what you are thinking': finding space for children's agency in the UK asylum system. *Area* 42(2): 162–169.

Cumbers, A., Helms, G. and Swanson, K. (2010) Class, agency and resistance in the old industrial city. *Antipode* 42: 46–73.

Daily Express (2002) 'Our town's too nice for refugees … they will try to escape, rapists and thieves will terrorise us'. 23 March 2002, p. 1.

Daily Express (2005) 'Bombers are all sponging asylum seekers'. 27 July 2005, p. 1.

Daily Star (2005) 'Asylum seekers ate our donkeys'. 21 August 2005, p. 1.

Darling, J. (2011a) Domopolitics, governmentality and the regulation of asylum accommodation. *Political Geography* 30: 263–271.

Darling, J. (2011b) Giving space: care, generosity and belonging in a UK asylum drop-in centre. *Geoforum* 42(4): 408–417.

Darling, J. (2014) Another letter from the Home Office: reading the material politics of asylum. *Environment and Planning D: Society and Space* 32(3): 484–500.

Davidson, J., Smith, M., Bondi, L. and Probyn, E. (2008) Emotion, space and society: Editorial introduction. *Emotion, Space and Society* 1(1): 1–3.

Day, K. and White, P. (2001) Choice or circumstance: The UK as the location of asylum applications by Bosnian and Somali refugees. *GeoJournal* 56(1): 15–26.

de Certeau, M. (1984) *The Practice of Everyday Life*. Oakland, CA: University of California Press.

Department for Communities and Local Government (2014) IMD 2010: postcodes and IMD rankings, bulk extracts for admin areas. Available at http://dclgapps. communities.gov.uk/imd/imd-by-postcode.html (accessed 15 December 2014).

Detention Action (2013) The financial waste of long term detention. London: Detention Action. Available at http://detentionaction.org.uk/wordpress/wp-content/uploads/ 2011/10/costs-briefing-03131.docx (accessed 29 July 2014).

Devlin, P. (1979) *The Judge*. Oxford: Oxford University Press.

Dickens, C. (1852–1853/1993) *Bleak House*. Wordsworth Editions.

Dikeç, M., Clark, N. and Barnett, C. (2009) Extending hospitality: giving space, taking time. In Dikeç, M., Clark, N. and Barnett, C. (eds) *Extending Hospitality: Giving Space, Taking Time*. Paragraph Special Issues, 32(1). Edinburgh: Edinburgh University Press, pp. 1–14.

Dixon, J., Durrheim, K. and Tredoux, C. (2005) Beyond the optimal contact strategy: a reality check for the contact hypothesis. *American Psychologist* 60: 697–711.

Dostoyevsky, F. (2003) *The Brothers Karamazov*. London: Penguin.

Doward, J. and Townsend, M. (2006) 'I will help you', he said. Then he asked for sex. *The Observer*, 21 May. Available at http://www.theguardian.com/politics/2006/ may/21/immigration.immigrationandpublicservices2 (accessed 30 August 2015).

du Gay, P. (2000) *In Praise of Bureaucracy: Weber - Organization - Ethics*. Sage Publications.

Edkins, J. and Pin-Fat, V. (2005) Through the wire: relations of power and relations of violence. *Millennium - Journal of International Studies* 34(1): 1–26.

England, P. and Folbre, N. (1999) The cost of caring. *Annals of the American Academy of Political and Social Science* 561(1): 39–51.

Eurostat (2015) *Eurostat Statistics Explained: Asylum Statistics*. Luxembourg: Eurostat. Accessed at http://ec.europa.eu/eurostat/statistics-explained/index.php/Asylum_ statistics (accessed 27 August 2015).

Farmer, A. (2013) The impact of immigration detention on children. *Forced Migration Review* 44: 14–16.

Fassin, D. (2005) Compassion and repression: the moral economy of immigration policies in France. *Cultural Anthropology* 20(3): 362–387.

Feldman, G. (2008) Democracy or technocracy? Rethinking the policymaker as specific intellectual. *JPoX - Journal on Political Excellence* (pilot issue). Available at http://jpox.eu/static/bf_pdf/pdfoutput.php?cid=170 (accessed 13 February 2013).

Figley, C. (ed.) (1995) *Compassion Fatigue: Secondary Traumatic Stress Disorders from Treating the Traumatized.* New York: Brunner/Mazel.

Figley, C. (ed.) (2002) *Treating Compassion Fatigue.* Oxford: Routledge.

Finney, N. and Simpson, L. (2009) *Sleepwalking to Segregation?: Challenging Myths About Race and Migration.* Bristol: Policy Press.

Fischer, W.F. (1970) *Theories of Anxiety.* New York: Harper & Row.

Friedman, M. (1991) The practice of partiality. *Ethics* 101(4): 818–835.

Friedman, M. (1993) *What are Friends For? Feminist Perspectives on Personal Relationships and Moral Theory.* Ithaca, NY, and London: Cornell University Press.

Fordham, M., Stefanelli, J. and Eser, S. (2013) Immigration detention and the rule of law: safeguarding principles. Bingham Centre for the Rule of Law. London: British Institute of International and Comparative Law. Available at http://www.biicl.org/files/6559_immigration_detention_and_the_rol_-_web_version.pdf (accessed 18 March 2014).

Fortier, A.-M. (2007) Too close for comfort: loving thy neighbour and the management of multicultural intimacies. *Environment and Planning D: Society and Space* 25: 104–119.

Foucault, M. (2008) *The Birth of Biopolitics: Lectures at the Collège de France 1978–1979.* New York: Palgrave Macmillan.

G4S (2013a) A day in the life of a Detainee Custody Officer (In Country Escorting). London: G4S. Available at http://careers.g4s.com/2010/11/a-day-in-the-life-of-a-detainee-custody-officer-in-country-escorting/ (accessed 8 April 2014).

G4S (2013b) A day in the life of a Detainee Custody Officer (Overseas Escorting). London: G4S. Available at http://careers.g4s.com/2010/11/a-day-in-the-life-of-a-detainee-custody-officer-overseas-escorting/ (accessed 8 April 2014).

Garber, M. (2004) Compassion. In L. Berlant (ed.) *Compassion: The Culture and Politics of an Emotion.* New York and London: Routledge.

Garcia, J., Nasho, E.-K. and Peretz, L. (eds) (2006) Voices From Detention II: A collection of testimonies from immigration detainees in the United Kingdom and Australia in their own words. Oxford: Barbed Wire Britain. Available at http://closecampsfield.files.wordpress.com/2011/03/voicesfromdetentionii.pdf (accessed 29 July 2014).

Gibney, M. (2004) *The Ethics and Politics of Asylum: Liberal Democracy and the Response to Refugees.* Cambridge: Cambridge University Press.

Gibney, M. (2008) Asylum and the expansion of deportation in the United Kingdom. *Government & Opposition* 43(2): 146–167.

Gibney, M. and Hansen, R. (2003) Asylum policy in the West: past trends, future possibilities. *Discussion Paper No. 2003/68.* Tokyo: United Nations University, World Institute for Development Economics Research (WIDER).

Gill, M. (2012) Care and value at the end of life. *Poetics* 40(2): 118–132.

Gill, N. (2009) Whose 'No Borders'? Achieving no borders for the right reasons. *Refuge* 26(2): 107–120.

Gill, N. (2010) New state-theoretic approaches to asylum and refugee geographies. *Progress in Human Geography* 5(34): 626–645.

Gill, N. (2014) Forms that form. In N. Thrift, A. Tickell, S. Woolgar and W.H. Rupp (eds) *Globalization in Practice.* Oxford: Oxford University Press, pp. 231–235.

Gill, N., Conlon, D., Oeppen, C. and Tyler, I. (2012) Networks of Asylum Support in the UK and USA: a Handbook of Ideas, Strategies and Best Practice for

Asylum Support Groups in a Challenging Social and Economic Climate. Available at http://steedee.files.wordpress.com/2012/03/networks-of-asylum-support-print2.pdf (accessed 18 January 2013).

Gill, N., Conlon, D., Tyler, I. and Oeppen, C. (2014) The tactics of asylum and irregular migrant support groups: disrupting bodily, technological, and neoliberal strategies of control. *Annals of the Association of American Geographers* 104(2): 373–381.

Girma, M., Radice, S., Tsangarides, N. and Walter, N. (2014) *Detained: Women Asylum Seekers Locked up in the UK*. London: Women for Refugee Women.

Glover, J. (1977) *Causing Death and Saving Lives: The Moral Problems of Abortion, Infanticide, Suicide, Euthanasia, Capital Punishment, War and Other Life-or-death Choices*. Harmondsworth, Middlesex: Penguin.

Goffman, E. (1961) *Asylums: Essays on the Social Situation of Mental Patients and Other Inmates*. New York: Anchor.

Goffman, E. (1967) *Interaction Ritual*. New York: Anchor.

Good, A. (2007) *Anthropology and Expertise in the Asylum Courts*. Glasshouse.

Gorz, A. (1968) Reform and revolution. *Socialist Register* 5: 111–143.

Gregory, D. (2013) Theory of the drone 10: killing at a distance. *Geographical Imaginations: War, Space and Security*. Available at http://geographicalimaginations.com/2013/09/15/theory-of-the-drone-10-killing-at-a-distance/ (accessed 15 December 2014).

Griffiths, D., Sigona, N. and Zetter, R. (2004) Integration and dispersal in the UK. *Forced Migration Review* 23: 27–29.

Grossman, D. (2009) *On Killing: The Psychological Cost of Learning to Kill in War and Society*. Back Bay Books.

Guild, E. (2000) *European Community Law from a Migrant's Perspective*. The Hague/London/New York: Kluwer Law International.

Guild, E. (2002) The border abroad: visas and border controls. In K. Groenendijk, E. Guild and P. Minderhoud (eds) *In Search of Europe's Borders*. The Hague/London/New York: Kluwer International Law, pp. 87–105.

Guiraudon, V. (2000) European integration and migration policy: vertical policy making as venue shopping. *Journal of Common Market Studies* 38(2): 163–195.

Guiraudon, V. (2003) Before the EU border: remote control of the "huddled masses". In K. Groenendijk, E. Guild and P. Minderhoud (eds) *In Search of Europe's Borders*. The Hague/London/New York: Kluwer International Law, pp. 191–214.

Gupta, A. (2006) Blurred boundaries: The discourse of corruption, the culture of politics, and the imagined state. In A. Sharma and A. Gupta (eds) *The Anthropology of the State: A Reader*. Oxford: Blackwell, pp. 211–242.

Hall, A. (2012) *Border Watch: Cultures of Immigration, Detention and Control*. London: Pluto Press.

Hamblet, W. (2003) The geography of goodness: proximity's dilemma and the difficulties of moral response to the distant sufferer. *The Monist* 86(3): 355–366.

Hamblet, W. (2011) Moral distance. In D. Chatterjee (ed.) *Encyclopedia of Global Justice*. Springer, pp. 717–719.

Hamnett, C. (1997) 'The sleep of reason?' *Environment and Planning D: Society and Space* 15(2): 127–128.

Hansard HC (5 December 2005) Vol. 440 Col. 972W.

Hansard HC (9 January 2006) Vol. 441 Col. 374W.
Hansard HC (14 March 2006) Vol. 443 Col. 97WS.
Hansard HC (28 January 2013) Vol. 557 Col. 531W.
Hansard HL (4 February 2010) Vol. 717 Col. 67WA.
Hargreaves, S., Holmes, A. and Friedland, J. (2005) Charging failed asylum seekers for health care in the UK. *The Lancet* 365(9461): 732–733.
Harvey, D. (2000) *Spaces of Hope.* Berkeley, CA: University of California Press.
Hayter, T. (2004) *Open Borders: The Case Against Immigration Controls.* London: Pluto Press.
Hemming, P.J. (2011) Meaningful encounters? Religion and social cohesion in the English primary school. *Social & Cultural Geography* 12(1): 63–81.
Henderson, V.L. (2008) Is there hope for anger? The politics of spatializing and (re) producing an emotion. *Emotion, Space and Society* 1: 28–37.
Herlihy, J. and Turner, S. (2006) Should discrepant accounts given by asylum seekers be taken as proof of deceit? *Torture* 16(2): 81–92.
Hess, F. (1998) *Spinning Wheels: The Politics of Urban School Reform.* Washington, D.C.: The Brookings Institution Press.
Hewstone, M. (2003) Intergroup contact: panacea for prejudice? *The Psychologist* 16: 352–355.
Heyman, J. (1995) Putting power in the anthropology of bureaucracy: the immigration and naturalization service at the Mexico-United States border. *Current Anthropology* 36(2): 261–287.
Hiemstra, N. (2013) 'You don't even know where you are': chaotic geographies of US migrant detention and deportation. In D. Moran, N. Gill and D. Conlon (eds) *Carceral Spaces: Mobility and Agency in Imprisonment and Migrant Detention.* Farnham: Ashgate, pp. 57–76.
HM Chief Inspector of Prisons (2014) Report on an unannounced inspection of Harmondsworth Immigration Removal Centre, 5-16 August 2013. London: HM Inspectorate of Prisons. Available at http://www.justice.gov.uk/downloads/publications/inspectorate-reports/hmipris/immigration-removal-centre-inspections/harmondsworth/harmondsworth-2014.pdf (accessed 29 July 2014).
HM Inspectorate of Prisons (2002) An inspection of Campsfield House Immigration Removal Centre. London: HM Inspectorate of Prisons. Available at http://www.justice.gov.uk/downloads/publications/inspectorate-reports/hmipris/immigration-removal-centre-inspections/campsfield-house/campsfieldhouse02-rps.pdf (accessed 29 July 2014).
Hochschild, A.R. (1995) The culture of politics: traditional, postmodern, cold-modern, and warm-modern ideals of care. *Social Politics: International Studies in Gender, State & Society* 2(3): 331–346.
Holt, L. (2007) Growing up global: economic restructuring and children's everyday lives (Review). *Environment and Planning A* 39(5): 1269–1270.
Home Affairs Select Committee (2013a) *Seventh Report: Asylum.* London: House of Commons. Available at http://www.publications.parliament.uk/pa/cm201314/cmselect/cmhaff/71/7102.htm (accessed 15 January 2014).
Home Affairs Select Committee (2013b) *Written Evidence: Asylum.* London: House of Commons. Available at http://www.publications.parliament.uk/pa/cm201314/cmselect/cmhaff/71/71we01.htm and http://www.publications.parliament.uk/pa/cm201314/cmselect/cmhaff/71/71vw01.htm (both accessed 30 August 2015).

Home Office (1998) *Fairer, Faster, Firmer: A Modern Approach to Immigration and Asylum.* Cm. 4018. London: HM Stationery Office. Available at https://www.gov.uk/government/uploads/system/uploads/attachment_data/file/264150/4018.pdf (accessed 29 July 2014).

Home Office (2004) *Review of Resourcing & Management of Immigration Enforcement: Final Report.* C. Pelham. London: Home Office.

Home Office (2006) *Controlling Our Borders: Making Immigration Work for Britain.* Five Year Strategy for Asylum and Immigration. Norwich: HM Stationery Office. Available at https://www.gov.uk/government/uploads/system/uploads/attachment_data/file/251091/6472.pdf (accessed 30 August 2015).

Home Office (2011a) Detention data tables: Immigration Statistics April to June 2011. Available at http://www.homeoffice.gov.uk/publications/science-research-statistics/research-statistics/immigration-asylum-research/immigration-tabs-q2-2011v2/detention-q2-11-tabs (accessed 29 July 2014).

Home Office (2011b) Policy Bulletin: Section 55 Guidance. London: Home Office. Available at https://www.gov.uk/government/uploads/system/uploads/attachment_data/file/257491/pb75.pdf (accessed 2 July 2014).

Home Office (2014a) Asylum Policy Instruction: Permission to Work. London: Home Office. Available at https://www.gov.uk/government/uploads/system/uploads/attachment_data/file/299415/Permission_to_Work_Asy_v6_0.pdf (accessed 2 July 2014).

Home Office (2014b) Asylum support. London: Home Office. Available at https://www.gov.uk/asylum-support/further-information (accessed 15 December 2014).

Horton, J. and Kraftl, P. (2009) What (else) matters? Policy contexts, emotional geographies. *Environment and Planning A* 41(12): 2984–3002.

Hubbard, P. (2005) Accommodating otherness: Anti-asylum centre protest and the maintenance of white privilege. *Transactions of the Institute of British Geographers* 30(1): 52–65.

Hull, M. (2012) *Government of Paper: the Materiality of Bureaucracy in Urban Pakistan.* Oakland, CA: University of California Press.

Hume, D. (1739/1896) *A Treatise of Human Nature by David Hume, Reprinted from the Original Edition in Three Volumes and Edited, with an Analytical Index.* Oxford: Clarendon Press.

Hunter, S. (2012) Ordering differentiation: Reconfiguring governance as relational politics. *Journal of Psycho-Social Studies* 6(1).

Hutcheson, F. (1971) Inquiry into the original of our ideas of beauty and virtue [1726]. In B. Fabian (ed.) *Collected Works of Francis Hutcheson.* Hildesheim: George Olms Verlagsbuchhandlung.

Hyland, J. (2000) 58 Chinese migrants found dead in lorry at Dover, Britain. World Socialist Web Site. Oak Park, MI: International Committee of the Fourth International. Available at http://www.wsws.org/en/articles/2000/06/immi-j21.html (accessed 29 July 2014).

Hyndman, J. and Giles, W. (2011) Waiting for what? The feminization of asylum in protracted situations. *Gender, Place & Culture* 18(3): 361–379.

Hynes, P. (2009) Contemporary compulsory dispersal and the absence of space for the restoration of trust. *Journal of Refugee Studies* 22(1): 97–121.

Immigration and Nationality Directorate (2001) Detention Centre Rules. SI 2001/238. London: HMSO. Available at http://www.legislation.gov.uk/uksi/2001/238/pdfs/uksi_20010238_en.pdf (accessed 18 March 2014).

Immigration and Nationality Directorate (2005) *The Service NASS Provides: How NASS Provides its Service.*

Independent Asylum Commission (2008a) *Fit for Purpose Yet?* The Independent Asylum Commission's Interim Findings. London: Citizen Organising Foundation.

Independent Asylum Commission (2008b) *Saving Sanctuary.* London: Citizen Organising Foundation.

Independent Chief Inspector of the UK Border Agency (2009) *Asylum: Getting the Balance Right?* London: Independent Chief Inspector of the UK Border Agency. Available at http://icinspector.independent.gov.uk/wp-content/uploads/2010/03/Asylum_Getting-the-Balance-Right_A-Thematic-Inspection.pdf (accessed 20 May 2013).

Jahoda, M. (1987) Contact and conflict in intergroup encounters. *British Journal of Psychology* 78: 275–277.

Jeffrey, A. (2013) *The Improvised State: Sovereignty, Performance and Agency in Dayton Bosnia.* Oxford: Wiley-Blackwell.

Jeffrey, C. (2010) *Timepass Youth, Class, and the Politics of Waiting in India.* Palo Alto, CA: Stanford University Press.

Jensen, N., Norredam, M., Priebe, S. and Krasnik, A. (2013) How do general practitioners experience providing care to refugees with mental health problems? A qualitative study from Denmark. *BMC Family Practice* 14(17).

Jessop, B. (2002) *The Future of the Capitalist State.* Cambridge: Polity.

Jones, R. (2007) *People/States/Territories.* Oxford: Blackwell.

Jones, R. (2012) State encounters. *Environment and Planning D: Society and Space* 30(5): 805–821.

Kant, I. (1788/2002) *Critique of Practical Reason.* Indianapolis: Hackett Publishing.

Karpf, A. (2002) We've been here before. *The Guardian*, 8 June. Available at http://www.theguardian.com/uk/2002/jun/08/immigration.immigrationandpublicservices (accessed 11 September 2015).

Katz, C. (2004) *Growing Up Global: Economic Restructuring and Children's Everyday Lives.* Minneapolis, MN: University of Minnesota Press.

Kelly, T. (2012) Soft-touch Britain, the asylum seeker capital of Europe: we let in more than anyone else last year. *Daily Mail*, 29 June. Available at http://www.dailymail.co.uk/news/article-2166738/Soft-touch-Britain-asylum-seeker-capital-Europe-We-let-year.html (accessed 11 September 2015).

Klein, N. (2007) *The Shock Doctrine: The Rise of Disaster Capitalism.* New York: Metropolitan Books.

Korf, B. (2007) Antinomies of generosity: moral geographies and post-tsunami aid in Southeast Asia. *Geoforum* 38(2): 366–378.

Koslowski, R. (2006) Towards an international regime for mobility and security? In K. Tamas and J. Palme (eds) *Globalizing Migration Regimes: New Challenges to Transnational Cooperation.* London: Ashgate, pp. 274–288.

Krznaric, R. (2014) *Empathy: A Handbook for Revolution.* London: Rider.

Lachs, J. (1981) *Responsibility of the Individual in Modern Society.* Brighton: Harvester.

Lagerwey, M.D. (2003) The nurses' trial at Hadamar and the ethical implications of health care values. In E. Baer and M. Goldenberg (eds) *Experience and Expression: Women, the Nazis, and the Holocaust.* Detroit, MI: Wayne State University Press, pp. 111–126.

Lahav, G. and Guiraudon, V. (2000) Comparative perspectives on border control: away from the border and outside the State. In P. Andreas and T. Snyder (eds) *The Wall Around the West: State Borders and Immigration Controls in North America and Europe*. Lanham, MD: Rowman and Littlefield, pp. 55–79.

Larner, W. and Craig, D. (2005) After neoliberalism? Community activism and local partnerships in Aotearoa New Zealand. *Antipode* 37(3): 402–424.

LaVan, M. (2003) Spaces of movement: performance, technology, transformation. *American Communication Journal* 6(3).

Lavenex, S. (2001) The Europeanization of refugee policies: normative challenges and institutional legacies. *Journal of Common Market Studies* 39(5): 851–874.

Lavenex, S. (2006) Shifting up and out: the foreign policy of European immigration control. *West European Politics* 29(2): 329–350.

Law, J. (2008) On sociology and STS. *Sociological Review* 56(4): 623–649.

Lefebvre, H. (1991) *The Production of Space*. Oxford: Blackwell.

Lefebvre, H. (2009) *State, Space, World: Selected Essays* (eds N. Brenner and S. Elden). Minneapolis, MN: University of Minnesota Press.

Leitner, H. (2012) Spaces of encounters: immigration, race, class, and the politics of belonging in small-town America. *Annals of the Association of American Geographers* 102(4): 828–846.

Leveson, B. (2012) *An Inquiry into the Culture, Practices and Ethics of the Press: Report*. London: The Stationery Office.

Levinas, E. (1979) *Totality and Infinity: An Essay on Exteriority*. New York: Springer.

Levinas, E. (1981) *Otherwise Than Being Or Beyond Essence*. New York: Springer.

Lewis, C.S. (2009) *The Screwtape Letters*. London: HarperOne.

Liberty (2014) Immigration detention. London: Liberty. Available at https://www.liberty-human-rights.org.uk/human-rights/asylum-and-borders/immigration-detention (accessed 23 July 2014).

Lipsky, M. (1980) *Street-Level Bureaucracy: Dilemmas of the Individual in Public Services*. New York: Russell Sage Foundation.

Macon, K.M. (2012) Bureaucratic regulation and emotional labor: Implications for social services case management. MA thesis, Department of Sociology. Atlanta, GA: Georgia State University.

Maiman, R. (2005) Asylum law practice in the United Kingdom after the Human Rights Act. In A. Sarat and S. Scheingold (eds) *The Worlds Cause Lawyers Make: Structure and Agency in Legal Practice*. Stanford University Press, pp. 410–424.

Malkki, L. (1996) Speechless emissaries: refugees, humanitarianism, and dehistoricization. *Cultural Anthropology* 11(3): 377–404.

Marsh, K., Venkatachalam, M. and Samanta, K. (2012) An economic analysis of alternatives to long-term detention. London: Matrix Evidence. Available at http://detentionaction.org.uk/wordpress/wp-content/uploads/2011/10/Matrix-Detention-Action-Economic-Analysis-0912.pdf (accessed 26 June 2014).

Martin, L. (2013) 'Getting in and getting out': legal geographies of US immigration detention. In D. Moran, N. Gill and D. Conlon (eds) *Carceral Spaces: Mobility and Agency in Imprisonment and Migrant Detention*. Aldershot: Ashgate, pp. 149–166.

Mason, R. (2013) Boris Johnson backs call for one-off amnesty for illegal immigrants. *The Telegraph*, 2 July 2013.

Massey, D. (2005) *For Space*. London: Sage.

Matejskova, T. and Leitner, H. (2011) Urban encounters with a difference: the contact hypothesis and immigrant integration projects in eastern Berlin. *Social & Cultural Geography* 12(7): 717–741.

Mbembe, J.-A. (2004) Aesthetics of superfluity. *Public Culture* 16(3): 373–405.

McGregor, J. (2011) Contestations and consequences of deportability: hunger strikes and the political agency of non-citizens. *Citizenship Studies* 15: 597–612.

McLaren, L.M. (2003) Anti-immigrant prejudice in Europe: contact, threat perceptions, and preferences for the exclusion of immigrants. *Social Forces* 81: 909–936.

McQueenie, K. (2005) Asylum queue disappears from Lunar House. *Croydon Guardian* 12 May 2005. Available at http://www.croydonguardian.co.uk/news/596013.print/ (accessed 18 December 2014).

Medic, N. (2004) *Making a Meal of a Myth*. Available at http://www.mediawise.org.uk/wp-content/uploads/2011/03/Making-a-meal-of-a-myth.pdf (accessed 28 July 2014).

Medical Justice (2008) *Outsourcing Abuse: The Use and Misuse of State-Sanctioned Force During the Detention and Removal of Asylum Seekers*. London: Medical Justice. Available at http://www.medicaljustice.org.uk/images/stories/reports/outsourcing%20abuse.pdf (accessed 26 June 2014).

Medical Justice (2010) '*State Sponsored Cruelty': Children in Immigration Detention*. London: Medical Justice. Available at http://www.medicaljustice.org.uk/images/stories/reports/sscfullreport.pdf (accessed 26 June 2014).

Medical Justice (2011) *Detained and Denied: The Clinical Care of Immigration Detainees Living with HIV*. London: Medical Justice. Available at http://www.medicaljustice.org.uk/images/stories/reports/d%26d.pdf (accessed 26 June 2014).

Medical Justice (2012) '*The Second Torture': The Immigration Detention of Torture Survivors*. London: Medical Justice. Available at http://www.medicaljustice.org.uk/reports-a-intelligence/mj/reports/2058-the-second-torture-the-immigration-detention-of-torture-survivors-22052012155.html (accessed 18 March 2014).

Medical Justice (2013) *Expecting Change: The Case for Ending the Immigration Detention of Pregnant Women*. London: Medical Justice. Available at http://www.medicaljustice.org.uk/images/stories/reports/expectingchange.pdf (accessed 26 June 2014).

Memon, A. (2012) Credibility of asylum claims: consistency and accuracy of autobiographical memory reports following trauma. *Applied Cognitive Psychology* 26(5): 677–679.

Merriam, S. (1998) *Qualitative Research and Case Study Applications in Education*. San Francisco, CA: Jossey-Bass.

Milgram, S. (1974/2005) *Obedience to Authority*. London: Pinter & Martin.

Moeller, S. (1999) *Compassion Fatigue: How the Media Sell Disease, Famine, War, and Death*. Abingdon: Psychology Press.

Mollard, C. (2001) *Asylum: The Truth Behind The Headlines*. Oxford: Oxfam Poverty Programme.

Morgan, M. (2011) *The Cambridge Introduction to Emmanuel Levinas*. Cambridge: Cambridge University Press.

Morrison, K. (2006) *Marx, Durkheim, Weber: Formations of Modern Social Thought*. London: Sage.

Mountz, A. (2010) *Seeking Asylum: Human Smuggling and Bureaucracy at the Border*. Minneapolis, MN: University of Minnesota Press.

Mountz, A. (2013) Mapping remote detention: dis/location through isolation. In J. Loyd, M. Mitchelson and A. Burridge (eds) *Beyond Walls and Cages: Prisons, Borders, And Global Crisis.* Athens, GA: The University of Georgia Press, pp. 91–104.

Mountz, A. and Hiemstra, N. (2014) Chaos and crisis: dissecting the spatiotemporal logics of contemporary migrations and state practices. *Annals of the Association of American Geographers* 104(2): 282–390.

National Association of Citizens Advice Bureaux (2002a) *Process Error - Asylum Support System Failing on all Fronts.* London: National Association of Citizens Advice Bureaux.

National Association of Citizens Advice Bureaux (2002b) *Distant Voices: CAB Clients' Experience of Continuing Problems with the National Asylum Support Service.* London: National Association of Citizens Advice Bureaux.

National Asylum Support Service (2005) Regionalisation Project Newsletter: August 2005 issue. Croydon: National Asylum Support Service.

National Asylum Support Service (2007) *Diagram of the 'End to End' Process of NASS Support.* Croydon: National Asylum Support Service.

National Audit Office (2007) The cancellation of Bicester Accommodation Centre. Home Office. Available at http://www.nao.org.uk/publications/0708/the_cancellation_ of_bicester_a.aspx (accessed 22 November 2013).

Nethery, A., Rafferty-Brown, B. and Yaylor, S. (2013) Exporting detention: Australia-funded immigration detention in Indonesia. *Journal of Refugee Studies* 26(1): 88–109.

Neumayer, E. (2006) Unequal access to foreign spaces: how states use visa restrictions to regulate mobility in a globalized world. *Transactions of the Institute of British Geographers* 31: 72–84.

Nietzsche, F. (1892/1961) *Thus Spoke Zarathustra: A Book for Everyone and No One.* Harmondsworth, Middlesex: Penguin Books.

Noble, G., Barnish, A., Finch, E. and Griffith, D. (2004) A review of the operation of the National Asylum Support Service. National Asylum Support Service. Croydon: Home Office.

No One Is Illegal (2003) *Manifesto.* Bolton: No One is Illegal. Available at http:// www.noii.org.uk/no-one-is-illegal-manifesto/ (accessed 16 July 2014).

North East Consortium for Asylum Support Services and North of England Refugee Service (2004) Asylum Seeking Communities in the North East of England: Quarterly Statistics as at March 2004. Newcastle-upon-Tyne: North of England Refugee Service. Available at http://www.refugee.org.uk/sites/default/ files/NASS%20regional%20stats%20March%2004.pdf (accessed 9 June 2014).

Nussbaum, M.C. (2001) *Upheavals of Thought: The Intelligence of Emotions.* Cambridge University Press.

Office for National Statistics (2014) Home page. Available at www.ons.gov.uk (accessed 15 December 2014).

Olle, N., Wallman, S., Grant, P., Bungey, S., Armstrong, P., Finger, S. and Martin, L. (2013) At work inside our detention centres: a guard's story. Available at http:// tgm-serco.patarmstrong.net.au (accessed 30 August 2015).

Owens, P. (2009) Reclaiming 'bare life'? Against Agamben on refugees. *International Relations* 23(4): 567–582.

Painter, J. (2006) Prosaic geographies of stateness. *Political Geography* 25: 752–774.

Paolini, S., Harwood, J. and Rubin, M. (2010) Negative intergroup contact makes group memberships salient: explaining why intergroup conflict endures. *Personality and Social Psychology Bulletin* 36: 1723–1738.

Patel, B. and Kelley, M. (2006) *The Social Care Needs of Refugees and Asylum Seekers*. London: Social Care Institute for Excellence.

Paveley, R. (2005) Finding God at Campsfield. *The Diocese of Oxford Reporter*, 2 November.

Peck, J. and Tickell, A. (2002) Neoliberalizing space. *Antipode* 34: 380–404.

Pérez, N.M. (2010) Emotions of queuing: a mirror of immigrants' social condition. In B. Sieben and Å. Wettergren (eds) *Emotionalizing Organizations and Organizing Emotions*. Basingstoke: Palgrave Macmillan, pp. 166–186.

Pettigrew, T.F. and Tropp, L.R. (2006) A meta-analysis test of inter-group contact theory. *Journal of Personality and Social Psychology* 90: 751–783.

Pettigrew, T.F. and Tropp, L.R. (2008) How does intergroup contact reduce prejudice? Meta-analytic tests of three mediators. *European Journal of Social Psychology* 32: 922–931.

Phelps, J. (2009) *Detained Lives: The Real Cost of Indefinite Immigration Detention*. London: London Detainee Support Group (now Detention Action). Available at http://detentionaction.org.uk/wordpress/wp-content/uploads/2011/10/Detained-Lives-report.pdf (accessed 30 August 2015).

Phillimore, J. and Goodson, L. (2006) Problem or opportunity? Asylum seekers, refugees, employment and social exclusion in deprived urban areas. *Urban Studies* 43(10): 1715–1736.

Phillips, D. (2006) Moving towards integration: the housing of asylum seekers and refugees in Britain. *Housing Studies* 21(4): 539–553.

Polese, C. (2013) Negotiating power between civil society and the state: the formulation of asylum policies in Italy and in the United Kingdom. PhD thesis. Department of Geography. London: University College London.

Popke, J. (2006) Geography and ethics: everyday mediations through care and consumption. *Progress in Human Geography* 30(4): 504–512.

Proctor, J. (1999) Introduction: overlapping terrains. In J. Proctor and D. Smith (eds) *Geography and Ethics: Journeys in a Moral Terrain*. London: Routledge.

Public and Commercial Services Union (2003) *Asylum Behind the Headlines*. London: Public and Commercial Services Union.

Public and Commercial Services Union (2014) Campaigns we support. London: Public and Commercial Services Union. Available at http://www.pcs.org.uk/en/campaigns/campaigns_we_support/index.cfm (accessed 5 August 2015).

Rachman, S. (1998) *Anxiety*. Hove: Psychology Press.

Rawlinson (2014) Private firms 'are using detained immigrants as cheap labour'. *The Guardian*, 22 August. Available at http://www.theguardian.com/uk-news/2014/aug/22/immigrants-cheap-labour-detention-centres-g4s-serco (accessed 11 September 2015).

Reader, S. (2003) Relationship and moral obligation. *The Monist* 86(3): 367–381.

Rhodes, R.A.W. (1994) The hollowing out of the state: the changing nature of the public service in Britain. *The Political Quarterly* 65(2): 138–151.

Rhodes, R.A.W. (1996) The new governance: governing without government. *Political Studies* XLIV: 652–667.

Rigo, E. (2007) *Europa di Confine: Trasformazioni della Cittadinaza nell'Unione Allargata.* Rome: Meltemi.

Robinson, V. and Segrott, J. (2002) *Understanding the Decision Making of Asylum Seekers.* Home Office Research Study 243. London: Home Office.

Robinson, V., Andersson, R. and Musterd, S. (2003) *Spreading the Burden? European Policies to Disperse Asylum Seekers.* Bristol: Policy Press.

Rock, P. (1993) *The Social World of an English Crown Court: Witnesses and Professionals in the Crown Court Centre at Wood Green.* Oxford: Clarendon Press.

Rodríguez-Pose, A. and Gill, N. (2003) The global trend towards devolution and its implications. *Environment and Planning C: Government and Policy* 21(3): 333–351.

Roman Catholic Diocese of Providence (2008) Bishop, pastors urge ICE to cease raids, allow agents to exercise 'conscientious objection'. Rhode Island: Roman Catholic Diocese of Providence. Available at http://www.diocesepvd.org/bishop-pastors-urge-ice-to-cease-raids-allow-agents-to-exercise-conscientious-objection/ (accessed 2 July 2014).

Rose, N. (1999) *Powers of Freedom: Reframing Political Thought.* Cambridge: Cambridge University Press.

Rothbart, M. and John, O.P. (1985) Social categorization and behavioral episodes: a cognitive analysis of the effects of intergroup contact. *Journal of Social Issues* 41(3): 81–104.

Sandel, M. (2009) *Justice: What's the Right Thing to Do?* London: Penguin.

Sayer, A. (2005) Class, moral worth and recognition. *Sociology* 39(5): 947–963.

Scheff, T. (1988) Shame and conformity: the deference-emotion system. *American Sociological Review* 53(3): 395–406.

Schuster, L. (2005) The realities of a new asylum paradigm. COMPAS Working Paper No. 20. Oxford: Oxford University Policy Documentation Centre. Available at http://www.compas.ox.ac.uk/publications/working-papers/wp-05-20/ (accessed 1 November 2013).

Schuster, L. and Majidi, N. (2013) What happens post-deportation? The experience of deported Afghans. *Migration Studies* 1(2): 221–240.

Scott, J. (1985) *Weapons of the Weak: Everyday Forms of Peasant Resistance.* New Haven and London: Yale University Press.

secondarytrauma.org (2010) What is secondary trauma? Aurora, CO: secondarytrauma.org. Available at http://secondarytrauma.org/contact.htm (accessed 23 July 2014).

Serres, M. (2007) *The Parasite.* Minneapolis, MN: University of Minnesota Press.

Sharma, A. and Gupta, A. (2006) *The Anthropology of the State: A Reader.* Oxford: Blackwell.

Sigona, N. (2010) Triple vulnerability: the lives of Britain's undocumented migrant children. OD 50:50. London: Open Democracy. Available at http://www.opendemocracy.net/5050/nando-sigona/triple-vulnerability-lives-of-britains-undocumented-migrant-children (accessed 6 June 2014).

Simanowitz, S. (2010) The body politic: the enduring power of the hunger strike. *Contemporary Review* 292: 1698.

Simmel, G. (1903/2002) The metropolis and mental life. In G. Bridge and S. Watson (eds) *The Blackwell City Reader.* Oxford: Wiley-Blackwell, pp. 103–110.

Singer, P. (1972) Famine, affluence and morality. *Philosophy and Public Affairs* 1(3): 229–243.

Smith, A. (1790) *The Theory of Moral Sentiments*. Library of Economics and Liberty. Available at http://www.econlib.org/library/Smith/smMS3.html (accessed 20 November 2014).

Smith, D. (1998) How far should we care? On the spatial scope of beneficience. *Progress in Human Geography* 22(1): 15–38.

Smith, D. (2000) *Moral Geographies: Ethics in a World of Difference*. Edinburgh: Edinburgh University Press.

Smith, F., Timbrell, H., Woolvin, M. *et al.* (2010) Enlivened geographies of volunteering: situated, embodied and emotional practices of voluntary action. *Scottish Geographical Journal* 126: 258–274.

Smith, K. (2015) Stories told by, for, and about women refugees: Engendering resistance. *ACME: An International E-Journal for Critical Geographies*, 2015, 14(2), 461–469.

Smith, M., Davidson, J., Cameron, L. and Bondi, L. (2009) Introduction – geography and emotion – emerging constellations. In M. Smith, J. Davidson, L. Cameron and L. Bondi (eds) *Emotion, Place and Culture*. Farnham: Ashgate.

South London Citizens (2009) South London Citizens Newsletter, Winter 2009. London: South London Citizens. Available at https://www.academia.edu/3666590/South_London_Citizens_Newsletter_Winter_2009 (accessed 2 July 2014).

Sparke, M. (2008) Political geography: political geographies of globalization III – resistance. *Progress in Human Geography* 32: 1–18.

Spivak, G. (1988) Can the subaltern speak? In C. Nelson and L. Grossberg (eds) *Marxism and the Interpretation of Culture*. Urbana, IL: University of Illinois Press, pp. 271–333.

Squire, V. (2009) *The Exclusionary Politics of Asylum*. Basingstoke: Palgrave Macmillan.

Stansfeld, W. (2013) UK cruelty towards asylum seekers and what should be done about it. San Francisco, CA: Scribd. Available at http://www.scribd.com/doc/125752329/UK-Cruelty-Towards-Asylum-Seekers-and-What-Should-Be-Done-About-It-v-1-4-2 (accessed 18 March 2014).

Stephan, W.G. and Stephan, C.W. (1985) Intergroup anxiety. *Journal of Social Issues* 41: 157–175.

Strauss, K. (2013) Unfree again: social reproduction, flexible labour markets and the resurgence of gang labour in the UK. *Antipode* 45(1): 180–197.

Taylor, D. and Mason, R. (2014) Home Office staff rewarded with gift vouchers for fighting off asylum cases. *The Guardian*, 14 January. Available at http://www.theguardian.com/uk-news/2014/jan/14/home-office-asylum-seekers-gift-vouchers (accessed 30 August 2015).

Taylor, D. and Muir, H. (2006) Red Cross aids failed asylum seekers. *The Guardian*, 9 January. Available at http://www.theguardian.com/news/2006/jan/09/immigration asylumandrefugees.uknews (accessed 30 August 2015).

Taylor, D. and Muir, H. (2010) Border staff humiliate and trick asylum seekers – whistleblower. *The Guardian*, 2 February. Available at http://www.theguardian.com/uk/2010/feb/02/border-staff-asylum-seekers-whistleblower (accessed 30 August 2015).

The Independent Chief Inspector of Borders and Immigration and Her Majesty's Inspectorate of Prisons (2012) *The Effectiveness and Impact of Immigration Detention*

Casework: A Joint Thematic Review by HM Inspectorate of Prisons and the Independent Chief Inspector of Borders and Immigration. London: HM Inspectorate of Prisons and The Independent Chief Inspector of Borders and Immigration. Available at http://icinspector.independent.gov.uk/wp-content/uploads/2012/12/Immigration-detention-casework-2012-FINAL.pdf (accessed 29 July 2013).

The Migration Observatory at the University of Oxford (2014) Asylum applications and estimated inflows, 1987–2011. Oxford: Migration Observatory at the University of Oxford. Available at http://www.migrationobservatory.ox.ac.uk/data-and-resources/charts/asylum-applications-and-estimated-inflows-1987-2011 (accessed 9 June 2014).

The Migration Observatory at the University of Oxford (2015a) Immigration detention in the UK. Updated by S. Silverman and R. Hajela. Available at http://migrationobservatory.ox.ac.uk/briefings/immigration-detention-uk (accessed 30 August 2015).

The Migration Observatory at the University of Oxford (2015b) Migration to the UK: Asylum, by Scott Blinder. Accessed at http://migrationobservatory.ox.ac.uk/briefings/migration-uk-asylum (accessed 27th August 2015).

The Secretary of State for the Home Department vs. Wayoka Limbuela and Binyam Tefera Tesema and Yusif Adam (2004) EWCA Civ 540. Case No: C/2004/0383, C2/2004/0384 & C/2004/0277.

The Sun (2003) 'Swan Baked … Asylum seekers are stealing and eating swans'. 4 July 2003, p. 1.

The Sun (2001) 'Has Hague abandoned all hope of winning this time?' 7 March 2001, p. 8.

Thomas, R. (2011) *Administrative Justice and Asylum Appeals: A Study of Tribunal Adjudication.* Oxford: Hart.

Thompson, J. (2004) Outrage over asylum centre. *Bristol Evening Post*, 21 April, p. 1.

Torre, M.E., Fine, M., with Alexander, N. et al. (2008) Participatory action research in the contact zone. In J. Cammarota and M. Fine (eds) *Revolutionizing Education: Youth Participatory Action Research in Motion.* New York: Routledge, pp. 23–44.

Tronto, J. (1987) Beyond gender difference to a theory of care. *Signs* 12(4): 644–663.

Tronto, J.C. (1993) *Moral Boundaries: A Political Argument for an Ethic of Care.* Oxford: Psychology Press.

Trouillot, M.-R. (2001) The anthropology of the state in the age of globalisation. *Current Anthropology* 42(1): 125–138.

Tuan, Y.-F. (1999) Geography and evil: a sketch. In J. Proctor and D. Smith (eds) *Geography and Ethics: Journeys in a Moral Terrain.* London and New York: Routledge, pp. 106–119.

Tutu, N. (ed.) (1989) *The Words of Desmond Tutu.* London: Hodder & Stoughton Religious.

Tyler, I. (2013) *Revolting Subjects: Social Abjection & Resistance in Neoliberal Britain.* London: Zed Books.

Tyler, I., Gill, N., Conlon, D. and Oeppen, C. (2014) The business of child detention: charitable co-option, migrant advocacy and activist outrage. *Race & Class* 56(1): 3–21.

UK Border Agency (2007) *Conducting the Asylum Interview.* London: United Kingdom Border Agency.

UK Border Agency (2011) Detainee custody officer certification. London: Home Office. Available at https://www.gov.uk/government/uploads/system/uploads/attachment_data/file/257734/detainee-custody-officer-cert.pdf (accessed 29 July 2014).

UNHCR (2002) 2002 UNHCR Statistical Yearbook – Chapter 1: Population Levels and Trends. London and Geneva: United Nations High Commission for Refugees. Accessed at: http://www.unhcr.org/41206f762.html (accessed 27 August 2015).

UNHCR (2005) Quality Initiative Project Second Report to the Minister. London and Geneva: United Nations High Commission for Refugees.

UNHCR (2006a) Quality Initiative Report: Third Report to the Minister. London and Geneva: United Nations High Commission for Refugees.

UNHCR (2006b) Quality Initiative Report: Key Observations and Recommendations. London and Geneva: United Nations High Commission for Refugees.

UNHCR (2007) Quality Initiative Project: Fourth Report to the Minister. London and Geneva: United Nations High Commission for Refugees.

UNHCR (2008) Quality Initiative Project Fifth Report to the Minister. London and Geneva: United Nations High Commission for Refugees.

UNHCR (2009) Quality Initiative Project: Key Observations and Recommendations. London and Geneva: United Nations High Commission for Refugees.

UNHCR (2014) UNHCR urges focus on saving lives as 2014 boat people numbers near 350,000. Geneva: United Nations High Commission for Refugees. Accessed at http://www.unhcr.org/5486e6b56.html (accessed 27 August 2015).

UNHCR (2015) UNHCR Global Trends: Forced Displacement in 2014. Geneva: United Nations High Commission for Refugees. Available at http://reliefweb.int/sites/reliefweb.int/files/resources/556725e69.pdf (accessed 27 August 2015).

UNITED (2015) List of 22,394 documented deaths of asylum seekers, refugees and migrants due to the restrictive policies of Fortress Europe. Amsterdam: UNITED for Intercultural Action. Accessed at http://www.unitedagainstracism.org/wp-content/uploads/2015/06/Listofdeaths22394June15.pdf (accessed 27 August 2015).

Urry, J. (2007) Mobilities. Policy Press.

Valentine, G. (1997) Tell me about…: using interviews as a research methodology. In R. Flowerdew and D. Martin (eds) Methods in Human Geography: A Guide for Students Doing a Research Project. Harlow: Longman, pp. 110–126.

Valentine, G. (2008) Living with difference: reflections on geographies of encounter. Progress in Human Geography 32(3): 323–337.

Valentine, G. (2010) Prejudice: rethinking geographies of oppression. Social & Cultural Geography 11(6): 519–537.

van Houtum, H. (2010) Human blacklisting: the global apartheid of the EU's external border regime. Environment and Planning D: Society and Space 28(6): 957–976.

Vaughan-Williams, N. (2009) The generalised bio-political border? Re-conceptualising the limits of sovereign power. Review of International Studies 35(4): 729–749.

Vickers, T. (2012) Refugees, Capitalism and the British State: Implications for Social Workers, Volunteers and Activists. Farnham, Surrey: Ashgate.

Waldron, J. (2003) Who is my neighbor? Humanity and proximity. The Monist 86(3): 333–354.

Watson, N., McKie, L., Hughes, B., Hopkins, D. and Gregory, S. (2004) (Inter)dependence, needs and care: the potential for disability and feminist theorists to develop an emancipatory model. Sociology 38(2): 331–350.

Webber, F. (2011) Does Barnardo's legitimise child detention? London: Institute of Race Relations. Available at http://www.irr.org.uk/news/does-barnardos-legitimise-child-detention/ (accessed 7 April 2014).

Webber, F. (2012) *Borderline Justice: The Fight for Refugee and Migrant Rights*. London: Pluto Press.

Weber, L. (2003) Down that wrong road: discretion in decisions to detain asylum seekers arriving at UK ports. *The Howard Journal* 42(3): 248–262.

Weber, L. and Bowling, B. (2002) The policing of immigration in the new world disorder. In P. Scraton (ed.) *Beyond September 11th: An Anthology of Dissent*. London: Pluto Press, pp. 123–129.

Weber, M. (1922/1987) Legitimate authority and bureaucracy. In L.E. Boone and D.D. Bowen (eds) *The Great Writings in Management and Organizational Behavior*. Boston: Irwin, pp. 5–19.

Weber, M. (1948) Politics as vocation. In M. Weber, H. Gerth and C.W. Mills (eds) *From Max Weber: Essays in Sociology*. London: Routledge, pp. 77–128.

Weizman, E. (2012). *The Least of All Possible Evils: Humanitarian Violence from Arendt to Gaza*. Verso Books.

Williams, R. (2012) Barnardo's chief: in the best interests of the children. *The Guardian*, 11 September. Available at http://www.theguardian.com/society/2012/sep/11/barnardos-chief-social-care-asylum-seekers (accessed 30 August 2015).

Wilson, H. (2013) Learning to think differently: diversity training and the 'good encounter'. *Geoforum* 45: 73–81.

Winder, R. (2013) *Bloody Foreigners: The Story of Immigration to Britain*. London: Abacus.

Woolley, A. (2014) BBC Radio 4 Four Thought: *Refugee Stories*, Series 4. Episode 33: broadcast 8 January 2014. Available on iPlayer Radio at http://www.bbc.co.uk/programmes/b03nt9wg (accessed 24 June 2014).

Yiu-leung, C.F. (2001) Snakeheads. 11 October 2001, Hong Kong: Martini [film]

Zelizer, V.A.R. (2005) *The Purchase Of Intimacy*. Princeton: Princeton University Press.

Zetter, R., Griffiths, D. and Sigona, N. (2005) Social capital or social exclusion? The impact of asylum-seeker dispersal on UK refugee community organizations. *Community Development Journal* 40(2): 169–181.

Zimmermann, S. (2014) Reconsidering the problem of 'bogus' refugees with 'socio-economic motivations' for seeking asylum. In N. Gill, J. Caletrio and V. Mason (eds) *Mobilities and Forced Migration*. London and New York: Routledge, pp. 35–52.

Žižek, S. (2009) *Violence: Six Sideways Reflections*. London: Profile Books.

Index

Nothing Personal?: Geographies of Governing and Activism in the British Asylum System,
First Edition. Nick Gill.
© 2016 John Wiley & Sons, Ltd. Published 2016 by John Wiley & Sons, Ltd.

2014 statistics 9–10
appeals against negative
  decisions 10–11, 14, 16, 19, 79–83,
  97–106, 183–5
bogus asylum seekers 2–3, 10, 50
crime/criminalisation 10–11, 12, 19,
  48–9, 63–4, 140–1, 170–1, 173–4
definition 2–3, 9–10, 17–18
dispersal 43, 49–55, 60, 65–6, 70–1
domestic violence 50, 55–6, 60–1, 92
economic migrants 2–3, 50, 151
English language difficulties 121–2
gay 122
healthcare issues 1, 4, 44, 48–9, 50–1,
  113–14, 119–24, 127–8, 141–2
interviews 79–83, 85–98, 164–5,
  183–4
non-belonging attitudes 124–5
paid work 125–6, 163–4, 177
perceptions 10–11, 48–52, 63–4, 88,
  109–10, 119–20, 141–2, 159–62,
  170–1, 186–8
social security benefits 10, 14,
  49–50, 52–65, 96, 105–6, 163–4,
  176, 182
stages of application 14, 54–5, 81–8,
  102–4
statistics 9–10, 53–5, 67, 69–70, 97–8,
  111–15, 118–19
terrorism 10, 18, 45, 131, 187
unfairness of accommodation-
  allocation concerns 88
in the United Kingdom 2–4, 9–15
asylum support groups 85–6, 131, 145–6,
  156–76, 177, 191–5
  see also support groups
audits 40–1

Bail for Immigration Detainees
  (BID) 113, 114
Barnardo's 148–50
Barnett, Clive 25–8, 35, 80, 102–3, 151,
  158
Bauman, Zygmunt 1, 4–5, 8, 16, 23, 27,
  33–8, 43–5, 51, 73, 130, 131, 135,
  137, 139, 146, 153–4, 185
Bayart, Jean-Francois 94, 186
BBC radio 143–4
BBC television 12, 90, 113
being-together-ness, potential of 30

belongings, immigration detention 124–6
Bethell, Nicholas 22
Bialasiewicz, Luiza 146–7
Bleak House (Dickens) 24
bogus asylum seekers 2–3, 10, 50
  see also economic migrants
  definition 2
  poverty of the notion 2–3, 50
Border and Immigration Agency 13
borders 5–8, 11–15, 23–4, 31–47,
  76–106, 140–55, 179–90
  see also state rescaling
  control systems 12–15, 31–47, 76–106,
    140–55, 179–90
  export (push-back) of borders 7, 41–2
Bosworth, Mary 118–20
Brenner, Neil 23, 37–8, 41
Bristol 51, 52, 58, 59, 62–4, 65, 66–8, 72,
  191
Bristol Refugee Inter-Agency Forum
  (BRIAF) 67
British National Party (BNP) 55, 74, 179
buffers 42, 51–2, 61, 65–9, 73,
  148, 183
  see also charities; mediators; third-sector
  organisations; volunteers
  conceptual discussion 51–2, 65–8, 73
bureaucracies 5–7, 8, 11, 13–14, 15–16,
  21–47, 48–75, 83–97, 118, 139–40,
  150, 160, 166–7, 182–90
  see also decision makers; governance;
  Lunar House; National Asylum
  Support Service; rationality
  conceptual discussion 5–7, 8, 13–14,
    15–16, 21–47, 48–75, 139–40,
    182–90
  definition 31–5
  dehumanisation 11, 33–5, 46,
    50–1, 118, 139, 150, 160, 166–7,
    181–90
  keeping people apart 30–5, 102–3
  mediators 5–6, 33–5, 45, 65–8
  moral distance 5–7, 8, 15–16, 21–47,
    48–75, 107–8, 139, 182–90
  obedience 27–8, 30–1
  objectives 31–5, 182–3
  Orwellian environment 84
  reform 187–90
  training 32, 83–8, 147–9
  Weber's theories 31–5